PRACTICAL
PLANT PHYSIOLOGY

PRACTICAL
PLANT PHYSIOLOGY

For

POSTGRADUATE AND UNDERGRADUATE [PASS AND HONOURS] STUDENTS

Dr M Acharyya Choudhuri

Reader in Botany, Burdwan University

Dr K K Gupta

Lecturer in Applied Botany, Burdwan University

New Central Book Agency (P) Ltd

LONDON

HYDERABAD ERNAKULAM BHUBANESWAR

NEW DELHI KOLKATA PUNE GUWAHATI

NCBA

REGD OFFICE

8/1 Chintamoni Das Lane, Kolkata 700 009, India
email: ncbapvtltd@eth.net

OVERSEAS

NCBA (UK) Ltd, 149 Park Avenue North, Northampton, NN3 2HY, UK
email: ncbauk@yahoo.com

EXPORT

NCBA Exports Pvt. Ltd, 212 Shahpur Jat, New Delhi 110 049
email: ncbaexports@ncbapvtltd.com

BRANCHES

2/27 Ansari Road, 1st Floor
Daryaganj, **New Delhi** 110 002
email: ncbadel@ncbapvtltd.com

House No. 3-1-315, 1st Floor
Nimboliadda, Kachiguda, **Hyderabad** 500 027
email: ncbahydb@ncbapvtltd.com

Shop Nos. 3 & 4, Vinayak Towers
681/B Budhwar Peth, Appa Balwant Chowk
Pune 411 002
email: ncbapune@ncbapvtltd.com

GSS Shopping Complex, 1st Floor
Opposite College Post Office, Convent Road
Ernakulam 682 035
0484-4021054 / 53

Shop No. 15, Bharati Towers, Ground Floor
Forest Park
Bhubaneswar 751 001
email: ncbabub@ncbapvtltd.com

Radhamayee Complex (Second Floor)
Opp. SBI, Panbazar, Jaswanta Road
Guwahati 781 001
email: ncbaguwahati@ncbapvtltd.com

PRACTICAL PLANT PHYSIOLOGY • Dr M Acharyya Choudhuri and Dr K K Gupta

First Published: 1976
Revised Reprint: 2009
Reprinted: 2017

PUBLISHER AND TYPESETTER
New Central Book Agency (P) Ltd
8/1 Chintamoni Das Lane, Kolkata 700 009

PRINTER
New Central Book Agency (P) Ltd
Web-Offset Division, Dhulagarh, Sankrail, Howrah

ISBN: 978 81 7381 554 6

Price ₹ 235.00

DEDICATED TO OUR REVERED PARENTS

Dr. S. K. Chatterjee,
Ph. D. (U.S.A.)

Professor and Head of the
Department of Botany
&
Dean, Faculty of Science.

THE UNIVERSITY OF BURDWAN
GOLAPBAG, BURDWAN
WEST BENGAL
Phone : BDN. 224—Extn. 81

Dated 12. 9. 1975.

FOREWORD

The study of life-processes in plants is an absorbing subject. The undersigned has been in it for the last 20 years or more and has come across only a few books which have dealt with how different processes connected with the life of plants can be studied in class. Some good books have been written by scientists of Western Countries and my experience is that these books do not always satisfy the needs of our students in correct perspective.

Dr. M. A. Choudhuri, Reader in Botany, and Dr. K. K. Gupta, Lecturer in Applied Botany of this University, have really satisfied me by their most wonderful treatments on the subject of their book entitled "Practical Plant Physiology". In an excellent way they have built up the story of life of plants, as seen in practice, for the students in undergraduate and post-graduate levels. I have least hesitation in strongly recommending the book to all the students of Plant Science, both in the disciplines of pure and applied, including agriculture. Teachers will find a valuable guidance in their treatment and the taught will find a clear idea how they should proceed to learn plant-life in laboratory.

I congratulate the writers for their dealing of the subject, particularly some chapters which will, undoubtedly, be of great help to both under-graduate and post-graduate students. I wish them hearty success in their bold attempt to write such a book which will fill up a big lacuna in the needs of students of Botany and Agriculture.

(S. K. Chatterjee)

PREFACE

Since their studenthood, the present authors have been quite aware of the dearth of a suitable book on practical Plant Physiology which could meet the need and the recurring demands of the students and the teachers of Colleges and Universities of our country. There are, of course, several books on the subject written by foreign authors, and according to logical expectation, much stress has been given in these books to those aspects which fulfil and satisfy the requirements of their own. Most of the experiments embodied in these books are suitable for performing in rather sophisticated and well-equipped laboratories which most of our science institutions do not possess. A few practical books on Botany written by Indian authors are, however, available now in the market but as most of them have dealt with all the branches of Botany, little weightage has been given to such a vast and dynamic discipline of Botany like Plant Physiology. In recent years, some good text books on the theoretical aspects of this branch of Botany have come out, but, unfortunately, the practical aspect still remains unattended to even by the veteran authors. Since it is the first venture of the present authors in this line, one might not think, rather unsympathetically, that "fools rush in where angels fear to tread." But one must admit that students as well as teachers very often face difficulties in negotiating practical classes for the want of a suitable book on practical Plant Physiology. We hope this alone is ample justification of our endeavour to undertake the writing of the present book in order to fulfil the long-awaited need and demand of the post-graduate and the under-graduate (pass and honours) students of Plant Physiology of all Indian Universities.

We have freely consulted the syllabi of different Indian Universities. It has been our deliberate intention to describe and discuss a sufficient number of experiments with illustrated figures and sketches, wherever necessary, to allow a considerable latitude of choice and to satisfy the need of M. Sc. and B. Sc. (pass and honours) courses of both the Botanical and Agricultural sciences. Careful attention has always been given to accommodate those experiments which will suit most of our laboratories. Each Chapter has begun with a general introduction which deals with the subject matter in sufficient details within the scope and purview of the present book. The experiments have been selected primarily to illustrate the basic facts and principles of Plant Physiology. A limited number of quantitative experiments have also been incorporated which are suitable for advanced students. The more simple experiments may be performed by the elementary students, while the remainder may be kept reserved for the more advanced syllabus. This book should also prove helpful to teachers of general Biology and general Botany, since many of the experiments included are sufficiently descriptive and simple for effective use as demonstrations in such courses.

At the end of the book, Appendixes have been included to describe some simple statistical calculations with suitable examples, preparation of reagents, use of log tables and many such useful informations and some relevant instructions to students which, we hope, will be very much helpful to students and teachers during day-to-day laboratory work and in evaluation of experimental results.

We are indebted to all who have contributed to the science of Plant Physiology. A number of experiments incorporated here are original, while the majority of them, though based on the investigations of other workers, have been suitably modified to serve the purpose of the present book. The authors will be amply rewarded if the book earns an appreciation from both the teachers and the taught. We are well aware of our limitations and shortcomings but we have spared no pains to present the book as a helpful guide for all concerned. Any constructive suggestion regarding the betterment of the book will be thankfully acknowledged.

The authors are deeply grateful to Dr. Subhendu Mukherjee, Lecturer in Botany, Calcutta University, for kindly going through the manuscript with much patience and also for his valuable suggestions. We are also grateful to ur teachers, Professor S. K. Chatterjee, Head of the Department of Botany, and Dean of the Faculty of Science of the University of Burdwan; Dr. C. T. Datta, Lecturer in Botany, Calcutta University; and Dr. S. K. Pain, Head of the Department of Botany, Viswa-Bharati, for their encouragement. We would like to record our sincere thanks to all of our friends and colleagues for their interest and encouragement. We are grateful to the Literary Executor of the late Sir Ronald A. Fisher, F.R.S., to Dr. Frank Yates, F.R.S. and to Longman Group Ltd., London, for permission to reprint Tables III, V and VII from their book *Statistical Tables for Biological, Agricultural and Medical Research* (*6th edition, 1974*).

We cannot thank our publisher Sri J. N. Sen enough, but for whose keen interest and kind intention the publication of this book would have hardly been possible. The authors are also thankful to Sri B. N. Roy for his kind co-operation in reading the proofs. Last but not the least, we are grateful to our artist Sri Gosaidas Pal for taking all the trouble of coming over to Burdwan in order to prepare figures and sketches for the book with painstaking efforts.

Janmastami, 1975.

M. A. C.

K. K. G.

CONTENTS

CHAPTER I

COLLOIDAL SYSTEMS

On the basis of researches on liquid diffusion, Thomas Graham (1861) divided some substances as *crystalloids*, e.g., all inorganic acids, bases and salts and many organic compounds, which could diffuse very rapidly in solution and could readily pass through animal and vegetable membranes; and some substances as *colloids*, e.g., starch, protoplasm, gelatine, proteins, etc., which failed to do so. The colloidal solution is a stable two-phase heterogeneous system, one of which (usually a solid), called *dispersed phase*, remains dispersed in a finely divided state throughout the other phase called the *dispersion phase* (generally a liquid). Among the colloidal systems, the liquid-in-liquid and the solid-in-liquid types are called *colloidal solutions*. The liquid-in-liquid type of colloidal solution is also called an *emulsoid solution.* A comparison of the size of the colloidal particles with ordinary molecules from the order of values is given in the following table.

Table 1. A comparison of the size of particles in true solutions, colloidal solutions and suspensions.

State of matter	Type of particles (dispersed phase)	Size of particles (dispersed phase)	Diffusibility of particles (dispersed phase)
True solutions	Molecules and ions	Less than 1 mμ	Will pass through membranes
Colloidal solutions	Molecular aggregates	1 to 200 mμ	Will not pass through membranes
Suspensions	Molecular aggregates	Over 200 mμ	Will not pass through membranes

1 mμ (milli micron) = 0·000001 mm (millimeter)

Colloidal system exhibiting properties of a fluid is known as *sol*. Sols are sometimes almost as thick as jelly, when they are called *gels*. Sols are subdivided into (a) *lyophobic* (solvent hating) and (b) *lyophilic* (solvent loving) colloids. Lyophobic sols (irreversible sols) are simple suspensions —so sometimes called *suspensoids* and here, there is no perceptible affinity between the dispersed phase and the dispersion phase (e.g., gold sol, silver sol, etc.). Lyophilic colloids (reversible) have, on the other hand, a marked affinity for solvent. This type of colloid is also called *emulsoids*. Most of the properties of a hydrophilic colloid may be explained by assuming that it is a hydrated suspensoid. This relationship is shown in Figure 1.

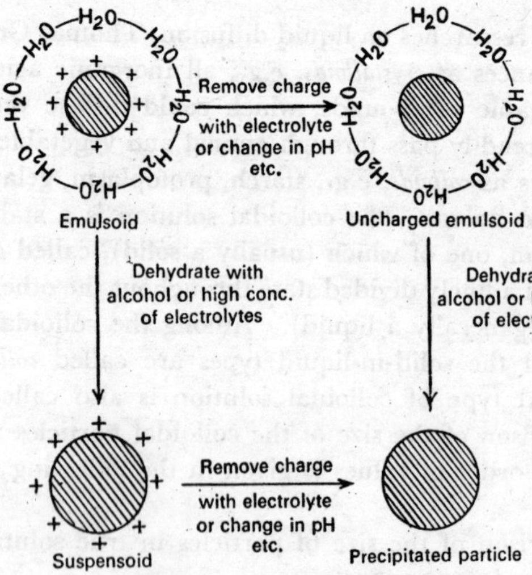

Fig. 1. Colloidal particles.

The protoplasm of plants consists of colloidal systems varying from emulsions (sols) to moderately stiff gels. The solid substances of protoplasm, proteins, dextrins, gums, etc., are characterised by the readiness with which they form colloidal systems in water. It is to these colloidal systems that protoplasm owes most of its characteristic properties, i.e., variation in viscosity from values not greater than that of pure water to those of jellies, pronounced capacity for imbibition of water, though itself high in water content, sometimes behaving as if immiscible with water; properties of irreversible coagulation in very high and low temperatures or to high concentration of salts, etc.

The experiments presented in this Chapter are intended only to illustrate a few of the basic principles of colloidal behaviour and to point out their possible relationships to the life processes of the plant.

1. **Preparation of suspensions**

(*i*) EXPERIMENT: A few ml of dil. H_2SO_4 is poured into a few ml of 5% $BaCl_2$ solution in a test tube. It is shaken well.

OBSERVATION: Precipitate of $BaSO_4$ formed gradually settles down forming suspensions.

INFERENCE: Particles of $BaSO_4$ being of larger dimensions (greater than 200 mμ) settle out gradually.

(*ii*) EXPERIMENT: A few pieces of charcoal are ground in a clean mortar. Three samples (0·2 gm each) of powdered charcoal are taken in three test tubes in which distilled water, 1% NH_4OH and 0·5% acetic acid solution are separately added, shaken and allowed to stand.

OBSERVATION: The rate of settling of charcoal powder is noted in each case. If they settle slowly, the effect of centrifuging is observed (at 500 r.p.m.).

INFERENCE: As in Expt. (i).

(*iii*) EXPERIMENT: 250 ml of distilled water is heated to boiling and 1 ml of 30% $FeCl_3$ solution is added slowly with constant stirring. An iron sol is thus formed.

OBSERVATION: The stability of a portion of iron sol is observed in concentrated HCl and NaOH solution.

INFERENCE: When $FeCl_3$ falls in water, it suffers hydrolysis, $FeCl_3 + 3\,H_2O = Fe(OH)_3 + 3\,HCl$. $Fe(OH)_3$ sol (it is actually a sol of hydrated ferric oxide, $Fe_2O_3\,(H_2O)_2$) is positively charged. Now, HCl being proton donor, $Fe(OH)_3$ remains in solution, whereas NaOH being proton acceptor, $Fe(OH)_3$ sol settles out.

2. Preparation of emulsions

(*i*) EXPERIMENT: A few drops of olive oil (or any other oil) are taken in a test tube. The tube is half-filled with absolute alcohol and shaken well. This is poured in a beaker of water.

OBSERVATION: A fine white emulsion of oil-in-water is formed.

INFERENCE: The particles of oil (dispersed phase) will not separate because the oil remains in such small droplets that a degree of stability is achieved. Here alcohol acts as a stabilizer or emulsifier.

(*ii*) EXPERIMENT: Equal quantities (1 ml) of olive oil are taken in two test tubes, and equal quantity of water is added to each. To one tube a few drops of 10% NaOH are added and shaken well.

OBSERVATION: An emulsion is formed in both the tubes. The oil separates out in the tube in which alkali has not been added while in the other tube the emulsion is stabilized.

INFERENCE: The shaking of oil with water does not necessarily make an emulsion, for, the oil and water soon separate into two layers when allowed to settle. For the stability of an emulsion the addition of an emulsifier is necessary. Here NaOH acts as an emulsifier and renders the emulsion permanent.

(*iii*) EXPERIMENT: A drop of oil-in-water emulsion is placed on a glass slide and a drop of oil is added at one side and gently mixed together. On another slide a drop of the oil-in-water emulsion and a drop of water are taken and mixed together. These are observed under the microscope. A drop of Sudan-III (0·1% in 50% ethanol) is added to each slide and again observed under the microscope.

OBSERVATION: In the first slide oil drop is not mixed with the emulsion while in the second slide water is mixed readily. In the first, red-stained larger oil globules are seen while in the second, fine red-stained particles are seen.

INFERENCE: In the first case water is in continuous phase, hence oil will not mix readily with this emulsion. In the second case water mixes immediately with the continuous phase (dispersion phase), i.e., water. This is clearly observed when stained with Sudan-III.

3. Preparation of colloids

(*i*) EXPERIMENT: 3 gm of soluble starch is taken in a beaker and dissolved in 100 ml water.

OBSERVATION: A colloidal solution of starch-in-water is formed.

INFERENCE: Here starch forms the dispersed phase of the colloidal system (being of dimension less than 200 mμ) in water which is the dispersion phase.

(*ii*) EXPERIMENT: A strong solution of sodium thiosulphate ($Na_2S_2O_3$) of approximately 3N strength is added drop by drop to conc. H_2SO_4. (Dissolve 372 gm $Na_2S_2O_3$, $5H_2O$ in 500 ml water in a volumetric flask to prepare 3N $Na_2S_2O_3$ soln.). This is cooled and mixed with water.

OBSERVATION: A colloidal solution of sulphur is formed.

INFERENCE: The solution on cooling deposits sulphur which is washed with water. This mass of sulphur dissolves completely in water to give a clear colloidal solution which may be purified by *dialysis* (the process of separating a crystalloid from a colloid by means of diffusion through a membrane).

4. Preparation of suspensoid sols (lyophobic) and their precipitation

(*i*) EXPERIMENT: 5 ml of 1% solution of $AgNO_3$ is taken in a 250 ml beaker and dil. NH_4OH is added until precipitate formed due to its addition, disappears. This is diluted with 100 ml of water. This solution is then mixed with equal volume of 0·5% tannic acid solution. (It is saved for the next experiment).

OBSERVATION: A colloidal solution of silver is formed.

INFERENCE: With excess NH_4OH, Ag_2O is produced which gets reduced to metallic silver on addition of tannic acid containing catechol and pyrogallol. Here metallic silver remains in dispersed phase.

(*ii*) EXPERIMENT: A few drops of $BaCl_2$ solution are added to the above colloidal solution.

OBSERVATION: The silver particles are precipitated.

INFERENCE: The particles of metallic silver Ag^- carry negative charge, hence they are readily precipitated by divalent cation Ba^{++} (or trivalent cation Al^{+++}, when $AlCl_3$ is added).

N.B. Very little electrolyte is necessary for this "salting out" of suspensoid sols, as these sols are very sensitive to traces of electrolytes.

5. Preparation of emulsoid sols (lyophilic) and their precipitation

(*i*) EXPERIMENT: 2 gm of dry starch is mixed well with distilled water and a paste is prepared. This paste is poured in 100 ml of boiling water in a beaker and boiled further for a few minutes with constant stirring. (The solution is saved for the next experiment).

OBSERVATION: An emulsoid sol of starch-in-water is obtained which remains unchanged when cooled.

INFERENCE: Starch being a lyophilic colloid absorbs large amount of water and, therefore, both the phases have the properties of liquids. The concentrated solution of starch is a dispersed phase while the weak solution of it is the dispersion phase. Thus, though starch is a solid, its colloidal solution may be regarded as an emulsoid sol.

(*ii*) EXPERIMENT: The emulsoid sol as prepared in the above experiment, is saturated with solid $(NH_4)_2SO_4$.

OBSERVATION: A white gelatinous precipitate is formed.

INFERENCE: Protein and starch are precipitated by salts because these form larger aggregates in concentrated salt solution. This is called "salting out" of protein or starch.

6. Preparation of gels and their precipitation

(*i*) EXPERIMENT: 1 gm of agar-agar is dissolved in 50 ml of water by boiling for half an hour and then cooled.

(The gel is saved for the next experiment).

OBSERVATION: The agar forms a thick opalescent solution (sol) which sets to a gel on cooling. This change is reversible.

INFERENCE: The sol of agar on cooling sets to a coherent jelly-like mass. This transparent or opalescent mass contains a large percentage of liquid which retains their shape and offers some resistance to mechanical deformation.

(*ii*) EXPERIMENT: The gel obtained from the previous experiment is warmed to form a sol which is taken in two test tubes and to one absolute alcohol and to the other saturated solution of $(NH_4)_2SO_4$ are added.

OBSERVATION: Agar mucilage is precipitated in each case.

INFERENCE: Precipitate consists simply of the aggregation of the smaller particles into larger ones which are big enough to settle down.

7. Preparation of reversible and irreversible colloids

(*i*) EXPERIMENT: To get a reversible colloid, Expt. 6 (*i*) is to be performed.

(*ii*) EXPERIMENT: The egg albumen is liquid at ordinary temperature but on heating it becomes solid.

OBSERVATION: The solid gel of egg albumen on cooling does not form a sol again.

INFERENCE: Egg albumen is an irreversible colloid. The solidification of the albumen on heating is due to coagulation of protein.

8. Properties of a colloidal solution

(*i*) BROWNIAN MOVEMENT OF COLLOIDS

(*a*) EXPERIMENT: A drop of sulphur sol from Expt. 3 (*ii*) is taken on a clean slide. It is then covered with a cover glass and observed with a beam of strong light under the high power of microscope.

OBSERVATION: The sulphur particles dispersed in water execute a ceaseless random motion.

INFERENCE: The random motion is due to the unbalanced impacts of the liquid molecules on the colloidal particles. The colloidal units are constantly hit from all sides by the surrounding solvent molecules and are tossed up. This erratic movement is called Brownian movement.

(*b*) EXPERIMENT: A drop or two of latex from plants (species of Apocynaceae, Euphorbiaceae, Asclepiadaceae, etc.) are taken, mounted on a slide and observed under the microscope.

OBSERVATION: The erratic zigzag movement of the colloidal particles (proteins, gums, etc.) are observed.

INFERENCE: As in Expt. 8 (*i.a*).

N.B. This experiment may also be performed with oil-water emulsion.

(*ii*) FLOCCULATION OF COLLOIDS

(*a*) FLOCCULATION OF HYDROPHOBIC SOLS:

EXPERIMENT: 5 ml of arsenious sulphide (AS_2S_3) sol is taken into each of three test tubes. A drop of 0·01 M NaCl (0·585 gm/litre) is added to the first tube, a drop of 0·01 M $CaCl_2$ (1·11 gm/litre) to the second, and a drop of 0·01 M $AlCl_3$ (1·335 gm/litre) to the third and observed immediately.

OBSERVATION: The sol is no longer stable and the particles are flocculated. The rate of flocculation is maximum with $AlCl_3$ and minimum with NaCl.

INFERENCE: The colloidal solution of arsenious sulphide has a negatively charged surface. Thus if NaCl, $CaCl_2$ and $AlCl_3$ solutions are added, Na^+, Ca^{++}, and Al^{+++} cations are preferentially adsorbed by the particles, the charge distribution on the surface of micelles of sol is disturbed resulting eventually in zero potential difference, when agglomeration of smaller particles into larger flakes takes place and the particles flocculate. Increasing valency has got increasing power of flocculation (valencies of Al, Ca and Na are 3, 2 and 1 respectively). $CaCl_2$ is about 88 times more efficient in flocculating As_2S_3 sol than NaCl and $AlCl_3$ is about 7 times more efficient in this respect.

(*b*) FLOCCULATION OF HYDROPHILIC SOLS:

EXPERIMENT: 2 gm of powdered gum acacia (gum arabic) is added slowly to 100 ml of boiling distilled water, stirring constantly until a homogeneous liquid results. 3 ml of gum acacia sol from the stock thus prepared is taken into 3 test tubes separately. To the first test tube 7 ml of 95% ethyl alcohol is added, shaken vigorously and observed. About half of the liquid in the tube is poured out and to the remainder about three times its volume of distilled water are added. To the second test tube 7 ml of 95% ethyl alcohol is added and shaken as before. Then about 0·2 gm of solid $AlCl_3$ is added, shaken until it dissolves and observed. To the third test tube 0·2 gm of solid $AlCl_3$ is first added, shaken and observed as in the preceding tubes. Then 7 ml of 95% ethyl alcohol is added and observed.

OBSERVATION: Lyophilic colloids are flocculated in all the test tubes. Precipitate of the first test tube is soluble in water.

INFERENCE: The gum arabic solution has negatively charged surface. $AlCl_3$ is a positively charged electrolyte and when these two oppositely charged colloidal particles are mixed, coagulation takes place which are hydrophobic. In the first test tube, the precipitated protein is a hydrophilic colloid which dissolves in water.

(*iii*) Isoelectric point of colloids

(*a*) Determination of isoelectric point of gelatin:

Experiment: 1 litre each of normal (approximately) solution of HCl (1 ml conc. HCl+11 ml water) and NaOH (40 gm/litre) is prepared and diluted to produce 500 ml each of 0·1 N, 0·01 N, 0·001 N and 0·0001 N. (Prepare the lower concentrations by dilution method as follows: If X be the desired strength and Y be the strength of the stock soln., take X ml of stock soln. and make the volume up to Y ml, i.e., add Y−X ml of water.). pH of each is determined using universal pH indicator paper. 2 gm of sheet gelatin is weighed into each of 11 beakers or wide-mouthed flasks and covered with 200 ml of solutions as follows:

Beaker No.	Solution	Beaker No.	Solution
1	1 N NaOH	7	0·0001 N HCl
2	0·1 N NaOH	8	0·001 N HCl
3	0·01 N NaOH	9	0·01 N HCl
4	0·001 N NaOH	10	0·1 N HCl
5	0·0001 N NaOH	11	1·0 N HCl
6	Distl. water		

The distilled water should be freshly boiled and cooled and this should be protected from atmospheric CO_2 in a stoppered wide-mouthed flask. The gelatin is allowed to soak 2 to 5 hr until the most hydrated sample begins to soften too much to handle. This is then filtered on folded cheese cloth or glass wool and the gelatin is weighed. The filtrates are saved and changes in pH are recorded. The gain in weight of the gelatin is plotted against the final pH of the medium and the approximate isoelectric point of the colloid is determined from the curve which corresponds to the pH at which minimum gain in weight takes place.

Observation: The change in weight of gelatin in each case is carefully noted.

Inference: The pH at which the minimum change in weight of gelatin due to absorption takes place gives the isoelectric point (round about 4·7 pH). The dispersion of many colloids in water (hydration) is affected by the hydrogen ion concentration or pH of the medium and is at minimum at a point called the isoelectric point. At isoelectric point the molecule is electrically balanced, i.e., in an electric field it will move neither to the cathode nor to the anode.

(*b*) Isoelectric point of plant tissue:

Experiment: 5 to 10 gm of potato discs is soaked for 24 hr. in the solutions prepared for Expt. 8 (*iii.a*). The changes in weight of the potato discs and the pH changes of the media are determined according to the previous experiment.

OBSERVATION and INFERENCE: As in Expt. 8 (*iii. a.*).

N.B. Isoelectric point of egg albumen may also be determined by adding 0·1 N NH_4OH and 0·1 N acetic acid separately and noting the pH at which maximum opalescence is produced.

(*iv*) IMBIBITION OF COLLOIDS

EXPERIMENT: Three lots of 10 or more grains of rice are weighed. First lot is covered with water, second with 7·5% NaCl and the third with 30% NaCl solutions for 24 to 72 hr. These lots are separately weighed. The volume of each lot is measured before and after treatments by displacement of water.

OBSERVATION: The changes in weights and volumes of grains of each lot are noted, and plotted on a graph paper.

INFERENCE: The colloids (protoplasm) of the seeds imbibe water osmotically and nonosmotically. In a solution of an electrolyte like NaCl, the nonosmotic active absorption forces are eliminated and water is drawn into the cells only osmotically and the absorption of water and solutes thus comes under general metabolic control of the tissues. Hence, imbibition of water is proportionally less in higher concentrations of NaCl.

(*v*) SURFACE TENSION

EXPERIMENT: The surface tension of the following liquids is measured arbitrarily by counting the time required by known volume (10 ml) of each liquid to fall completely from a burette. 10 ml each of the following liquids—ethanol, acetone, chloroform, ether, toluene, xylene, mercury and water are taken in a dry burette one by one and the time required by each liquid to fall completely through the burette is recorded with the help of a stop watch. Reading with each liquid may be taken at least three times.

RESULTS: The average time taken is noted in each case and entered in a table.

DISCUSSION: The attractive force between like molecules causing them to cling together is called *cohesion* while the similar attractive force between unlike molecules is called *adhesion.* The surface tension depends upon the force of adhesion between two unlike molecules. The greater the force of adhesion, the greater is the surface tension. In the present experiment it is found that the time taken by different liquids is different and if put in order on the basis of decreasing requirement of time, the liquids can be arranged as follows: (1) Mercury, (2) Water, (3) Xylene, (4) Toluene, (5) Chloroform, (6) Ether, (7) Acetone and (8) Ethanol. (The surface tension determines to some extent the shapes and distribution of protoplasmic structures and the more liquid parts of plant cells).

CHAPTER II

DIFFUSION, PERMEABILITY, OSMOSIS AND IMBIBITION

The plant cell has a rigid wall, internally bounded by a cell membrane, outside which lies the environment. Hence, the solid particles cannot get in or come out from the cell freely. To have an entry into or exit from the cell, the substances must be in a permeable form, i.e., either in a liquid or in a gaseous form. This movement of substances is governed by some physical processes like diffusion, permeability, osmosis, etc.

DIFFUSION: The movement of molecules of gases, liquids or solutes from the region of higher concentration to the region of lower concentration until the molecules are evenly distributed throughout the available space is known as *diffusion*. Thus diffusion may be said to be a movement of molecules from a region of high partial pressure to the region of low partial pressure as a result of their inherent kinetic energy. Various substances which enter into the plants from their surroundings out of which plants synthesize the numerous complex organic compounds, find their way inside by the process of diffusion.

Gases, liquids or solutes can diffuse simultaneously and independently at different rates and in different directions in the same place without interfering with each other. The rate of diffusion of gases is generally fast, while the liquids and solutes have a slower rate of diffusion. The rate of diffusion is directly proportional to the temperature and inversely proportional to the square root of density.

PERMEABILITY: A membrane which allows both water as well as solute particles to pass through is called *permeable*. The membrane which allows the passage of only the solvent particles and not the solute particles is called *semipermeable* or *differentially permeable*. *Impermeable membranes* are those which prohibit the entry of both solvent and solute particles. A plant cell normally has a differentially permeable membrane (the plasmamembrane). The permeability of the complex cell membrane depends on the nature of the surrounding particles as well as on the changing conditions inside and outside the cell. Among all substances, water has the most rapid rate of diffusion through cell membranes. The membrane is very much permeable to gases like CO_2, O_2, N_2, and fat-solvents like alcohol, ether, chloroform, etc., relatively less permeable to some organic substances like mono- and di-saccharides, fatty acids, amino acids, strong electrolytes, salts, bases, etc., and relatively impermeable to polysaccharides, phospholipids, proteins, etc.

Ions also affect the permeability of cytoplasmic membrane and there exists some sort of mutual active opposition between salts and their ions having the same electric charge. This phenomenon is known as 'antagonism of ions'. Permeability is increased by the monovalent Na^+ ions the effect of which is antagonised by the divalent Ca^{++} ions.

OSMOSIS: *Osmosis* is regarded as spontaneous diffusion of solvent across a semipermeable membrane from a place of its higher diffusion pressure (or concentration) to a place of its lower diffusion pressure (or concentration). The *osmotic pressure* is that excess pressure which has to be applied to the solution when it is separated from the solvent by means of a semipermeable membrane in order to check the inflow of the solvent and establish an equilibrium in the system. The osmotic pressure of a solution is, therefore, a rating of the potential maximum pressure which can be developed in a solution as a result of osmosis. This pressure is directly dependent upon the concentration of the solute particles as in nonelectrolytes. Hence molar solutions are generally used in experiments of osmosis which takes into account the total number of particles of the solutes. In the case of electrolytes, the solute dissociates into cations and anions in water, increasing the number of solute particles. Thus the osmotic pressure of a molar solution of an electrolyte will be twice as much as that of a nonelectrolyte when completely dissociated.

The pressure which develops in a cell from time to time due to osmotic diffusion of water is called the *turgour pressure* (or the *hydrostatic pressure*). The wall will exert a reverse pressure on cell solution. This is *wall pressure* which is always equal and opposite to turgour pressure. The turgour pressure in a particular system may be variable whereas osmotic pressure of that system is always constant. During endosmosis, the turgour pressure increases gradually in a solution and when it has reached its maximum limit it is equal to the osmotic pressure, provided the solvent is pure. In case of plant cells the actual value of osmotic pressure cannot be obtained because of its increasing cell volume due to endosmosis (cell wall is not perfectly inelastic). The *suction pressure* or the *diffusion pressure deficit* of water in a solution is that pressure which is the differencne of diffusion pressure of water in pure state and in solution, both under same temperature and pressure. The following fundamental relation between the three osmotic quantities of a plant cell can be expressed in simple equations,

Suction pressure (S) = Osmotic pressure (P) − Turgour Pressure (T).

or, $S = P - T$.

An ideal molar solution should have an osmotic pressure of 22·4 atmospheres at 0°C. If *n* gram molecules (molecular weight expressed in grams) of solutes are present in 1 litre of water, the diffusion pressure of water in the solution formed will be less than the diffusion pressure of pure water by $n \times 22{\cdot}4$ atmospheres. At N.T.P. the diffusion pressure of pure water is about 1240 atmospheres. Solutions having equal osmotic pressure or concentration are called *isotonic* or *isosmotic*. When a solution is at a higher concentration than the other, it is called *hypertonic* and the latter solution having lower concentration is called *hypotonic*. When a plant cell is immersed in a hypertonic solution exosmosis results and a shrinkage of protoplasm takes place. This phenomenon is known as *plasmolysis*. When plasmolysis is just to commence (i.e., turgour is zero), the cell is said to be at *incipient plasmolysis*. If a plasmolysed cell is transferred from a hypertonic solution to pure water, it slowly recovers by endosmosis of water regaining its turgour. This phenomenon is known as *deplasmolysis*.

In living cells many kinds of selectively permeable membranes are found such as plasmalemma, tonoplast, nuclear membrane, etc. The cellulose cell wall is, however, permeable to most type of substances including water. These cell walls when impregnated with cutin or suberin or waxy materials become impermeable to water.

IMBIBITION: Many substances (lyophilic) when immersed in certain solvents or solutions imbibe the solvent into their structure forming a colloidal solution. This taking up of water and the consequent swelling of colloidal materials and other substances is called *imbibition*. Substances such as proteins, gums, mucilage, starch, gelatin, pectin, agar-agar, cellulose, etc., form a viscous colloidal solution in water by the process of imbibition. Seeds contain substances which can imbibe water and the germination process is initiated. This imbibition of water by dispersed particles in a colloidal system is responsible for a number of physiological processes on the surface of protoplasm.

Different organic substances have various degrees of imbibing ability. For example, proteins are better imbibant than starch and cellulose. The amount of liquid imbibed depends upon the amount of mutual cohesive attraction between the molecules of the imbibant as well as on the amount of adhesive force between the molecules of imbibant and the imbibed liquid. The rate of imbibition depends on the diffusion pressure gradient of the solvents of the imbibant and the medium. The physical conditions of the medium such as pH, temperature, concentration, etc., also determine the rate of imbibition and the amount of liquid imbibed. Increase in temperature, within certain limits, increases the rate of imbibition. During imbibition energy may be released causing change in temperature. By imbibition the imbibant swells in volume producing an outwardly directed pressure called the *imbibition pressure* (analogous to turgour pressure). This pressure depends upon the nature of imbibant and the imbibed liquid, concentration, temperature, pH, etc.

1. Experiments on diffusion

(i) Diffusion of solid in liquid

Experiment: A beaker is almost filled with water. Some crystals of $CuSO_4$ or $KMnO_4$ are dropped carefully without disturbing water and is left as such for some time.

Observation: The water is uniformly coloured, blue in case of $CuSO_4$ and pink in case of $KMnO_4$.

Inference: The molecules of the chemicals diffuse gradually from higher concentration to lower concentration and are uniformly distributed after some time. Here, $CuSO_4$ or $KMnO_4$ diffuses independently of water and at the same time water diffuses independently of the chemicals.

(ii) Diffusion of liquid in liquid

Experiment: Two test tubes are taken. To one 30 mm depth of chloroform and to the other 4 mm depth of water are added. Now to the first test tube 4 mm depth of water and to the other 30 mm depth of ether are added (both chloroform and ether form the upper layer). Ether must be added carefully to avoid disturbance of water. The tubes are stoppered tightly with corks. The position of liquid layers in each test tube is marked and their thickness measured. The tubes are set aside for some time and the thickness of the liquids in each test tube is recorded at different intervals.

Observation: The rate of diffusion of ether is faster than that of chloroform into water as indicated by their respective volumes.

Inference: The rate of diffusion is inversely proportional (approximately) to the square root of density of the substance. Substances having higher molecular weights show slower diffusion rates than those having lower molecular weights. In the present experiment ether (C_2H_5–O–C_2H_5, mol. wt. 74) diffuses faster into water than chloroform ($CHCl_3$, mol. wt. 119·5). This ratio (74 : 119·5) is known as *diffusivity* or *coefficient of diffusion*.

(iii) Diffusion of gas in gas

Experiment: One gas jar is filled with CO_2 (either by laboratory method: $CaCO_3+HCl$, or by allowing living plant tissue to respire in a closed jar). Another jar is similarly filled with O_2 (either by laboratory method: MnO_2+KClO_3, or by allowing green plant tissue to photosynthesize in a closed jar). The gases may be tested with glowing match stick. The oxygen jar is then inverted over the mouth of the carbon dioxide jar and made air-tight with grease. It is then allowed to remain for some time. The jars are carefully removed and tested with glowing match stick.

Observation: The glowing match sticks flared up in both the jars.

INFERENCE: The diffusion of CO_2 and O_2 takes place in both the jars until finally the concentrations are same in both of them making a mixture of CO_2 and O_2. Hence the glowing match sticks flared up in both the jars.

(*iv*) COMPARATIVE RATES OF DIFFUSION OF DIFFERENT SOLUTES

EXPERIMENT: 3·2 gm of agar-agar is completely dissolved in 200 ml of boiling water and when partially cooled, 30 drops of methyl red soln. and a little of 0·1 N NaOH are added to give an alkaline yellow colour. 3 test tubes are filled three-fourth full with agar mixture and allowed to set. The agar is covered with 4 ml portion of the following solutions, stoppered tightly and kept in a cool place: (*a*) 4 ml of 0·4% methylene blue, (*b*) 4 ml of 0·05 N HCl, and (*c*) 2 ml of 0·1 N HCl plus 2 ml of 0·4% methylene blue. The diffusion of various solutes is recorded in millimeters after 4 hours. The top of the gel should be marked before the above solutions are added.

OBSERVATION: The rate of diffusion of HCl alone (tube b) is faster compared to the combination of methylene blue and HCl (tube c) and minimum in case of methylene blue alone (tube a).

INFERENCE: Different substances like gases, liquids and solutes can diffuse simultaneously and independently at different rates in the same place without interfering each other. HCl being gaseous in nature and of lower molecular weight can diffuse much faster than methylene blue which is a dye of higher molecular weight having an adsorptive property. Hence in combination, these two substances diffuse more readily than methylene blue alone.

(*v*) COMPARATIVE RATES OF DIFFUSION THROUGH DIFFERENT MEDIA

EXPERIMENT: Two apparatus for measuring comparative rates of diffusion through gas and liquid are set up (Figure 2). Tube 1 is filled with a 2% agar-sol containing 1 ml of methyl red indicator and 1 drop of 0·1 N NaOH (in alkaline medium methyl red is yellow and in acidic medium red). When the agar sets in the tube, it is held over a small bottle containing conc. HCl. The distance to the diffusion front as indicated by the red colour-

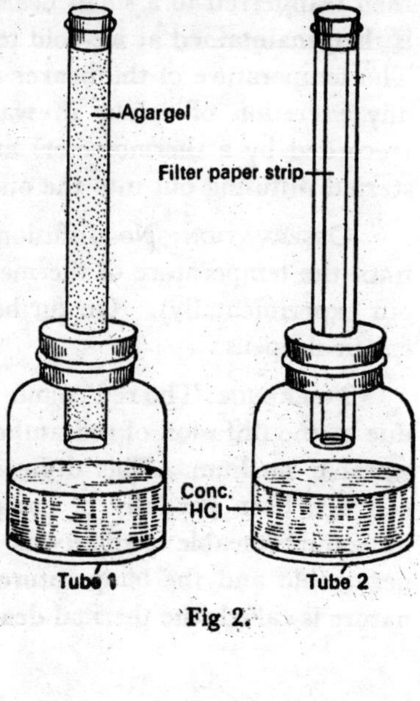

Fig 2.

line is measured at suitable time intervals and the rate of diffusion of HCl gas into agar-gel in millimeter per hour is recorded.

A strip of filter paper approximately of equal length and internal diameter of tube 2 is cut, soaked in methyl red indicator containing a little 0·1 N NaOH (the strip is coloured yellow) and suspended in tube 2. The tube is held over a small bottle containing conc. HCl as in tube 1. The rate of diffusion of HCl through the gaseous medium surrounding the strip is recorded by noting the colour change of the strip.

Observation: The rate of diffusion of HCl gas is faster in case of tube 2 containing filter paper strip and slower in case of tube 1 containing agar-gel.

Inference: The rate of diffusion of gases through a medium is inversely related to the density of the medium. Hence HCl diffuses faster in gaseous medium (tube 2) than in semisolid medium (tube 1).

N.B. The presence of the agar in tube 1 does not materially retard diffusion of solutes; the rate of diffusion through this gel may be taken, therefore, to be equal to the rate of diffusion through water.

2. Experiments on permeability

(*i*) Effect of temperature on permeability (thermal death point)

Experiment: 6 equal slices of beet root are cut with the help of a cork borer. These uniform slices are thoroughly washed in water to remove adhering anthocyanin pigment from the wounded cells. The slices are then transferred to a small beaker containing distilled water. The beaker is then maintained at ice-cold temperature and then at room temperature. The temperature of the beaker is then raised by 1°C. using hot water and any excretion of colour in water is carefully noted. The temperature (recorded by a thermometer) at which red anthocyanin pigment has just started diffusing out into the outer medium is cautiously recorded.

Observation: No diffusion of red anthocyanin pigment takes place until the temperature of the medium is raised to 'X' °C. (X is to be found out experimentally). On further increase of temperature the intensity of colour deepens.

Inference: The red colour of water at higher temperature is obviously due to the diffusion of red anthocyanin pigment from the cell sap into the external medium. This diffusion of pigment takes place only when the cell has been partially or completely killed at higher temperatures and its semipermeable property is lost. The cell membrane now becomes permeable and the temperature at which the cell loses its semipermeable nature is called 'the thermal death point'.

N.B. The thermal death point varies from species to species and depends on the physical conditions of the materials.

Chemicals like alcohol, copper sulphate, etc., also kill the cell destroying its semipermeable nature. At a particular conc. of these substances the excretion of red colour takes place indicating the destruction of semipermeable nature of the membrane.

(*ii*) Permeability of different dyes through cell membrane

Experiment: A solution of 0·05 M of $CaCl_2$ (5·55 gm/litre) is prepared. Equal quantities of methylene blue and neutral red dye solutions (100 mg dye in 100 ml of 50% alcohol) are then poured into two beakers containing equal quantity of $CaCl_2$ solution (the ratio of dye : $CaCl_2$ is 1 : 1000). Two clean slides are taken and to each slide one drop of neutral red-calcium chloride solution is taken. To one slide onion scales and to the other *Azolla* root hairs are placed. The time of penetration of the dye as indicated by the coloured cytoplasm is noted under the microscope. Similar experiment is done with methylene blue-calcium chloride solution and the time of penetration of the dye is also noted.

Observation: It is observed that the time of penetration in case of methylene blue is much longer than that of neutral red. (Compare the rate of penetration of dyes in the two materials).

Inference: The rate of entrance of different non-toxic dyes is different. This differential rate is due to the molecular arrangement of dyes, their adsorptive capacity and their solubility in water and lipid substance which is one of the important components of the cell membrane. The purpose of using $CaCl_2$ solution is to keep the physical characteristic of the membrane unaltered.

(*iii*) Experiments on differential permeability of a true solution and a colloidal solution

Experiment: 1% iodine solution is taken in a test tube and its mouth is covered with a semipermeable membrane (parchment or egg membrane may be used). The test tube is then inverted in a beaker containing 1% starch solution. In another test tube 1% starch solution is taken, covered with a semipermeable membrane and is inverted in another beaker containing 1% iodine solution (1 gm I_2 + 2 gm KI + 300 ml distilled water).

Observation: After about an hour the characteristic blue colour of starch-iodide complex appears in the beaker containing starch solution. But there is no change in colour in the beaker containing iodine solution. The test tube containing starch solution is also coloured blue.

Inference: The experiment clearly indicates that the molecules of iodine pass easily through the membrane in either direction while starch, which is a colloidal solution, does not.

N.B. To prepare 1% starch solution, 1 gm soluble starch is mixed with 100 ml water, warmed until the solution is clear. To avoid contamination, a pinch of salicylic acid should be added for preservation purpose.

(*iv*) IMPERMEABILITY OF SEED COAT

EXPERIMENT: 3 gm each of castor and gram seeds is taken and weighed. The seeds are then placed in two beakers containing equal quantity of water. After 1 hour seeds are taken out, blotted well and weighed again.

OBSERVATION: It is observed that no marked increase in weight takes place in case of castor seeds whereas considerable increase in weight takes place in case of gram seeds.

INFERENCE: The castor seeds having hard and waxy seed coat, are relatively impermeable to water. The gram seeds having no such physical barrier are quite permeable to water. Hence marked increase in weight in case of gram seeds occurs.

(*v*) EFFECT OF IONS ON PERMEABILITY AND THEIR ANTAGONISM

EXPERIMENT: Two stock solutions of NaCl and $CaCl_2$ (anhydrous) of conc. 1·88% (which is more or less isotonic to cell sap) are prepared. Six petridishes are taken and the solutions are added in the following proportions:

(a) Distilled water or demineralised water—(9 ml)
(b) NaCl : $CaCl_2$—1 : 0 (9 ml plus 0 ml)
(c) NaCl : $CaCl_2$—1 : 1 (4·5 ml plus 4·5 ml)
(d) NaCl : $CaCl_2$—2 : 1 (6 ml plus 3 ml)
(e) NaCl : $CaCl_2$—1 : 2 (3 ml plus 6 ml)
(f) NaCl : $CaCl_2$—0 : 1 (0 ml plus 9 ml)

Equal slices of beet root of uniform thickness are cut with the help of a cork borer and these are thoroughly washed in distilled water to remove the adhering anthocyanin pigment from the wounded cells. Four such slices are then transferred to each petridish containing the above solutions in the aforementioned proportions. Observations are made at 5 minutes interval for half an hour to note whether any red colour diffuses into the external solution.

OBSERVATION: Diffusion of red colour appears first in petridish (b), then in (c) and lastly in (a). No colour diffusion takes place in case of petridishes (d), (e) and (f).

INFERENCE: NaCl increases the permeability of the plasmamembrane to a point which is injurious to it but $CaCl_2$ acts in a reverse way, i.e., it decreases the permeability to a point which is injurious. When these two salt solutions are mixed in a proper ratio, their combined effect on the plasmamembrane neutralizes the individual effect and the permeability remains unaltered. Since Ca is a divalent cation (having two positive charges) and Na being a monovalent cation (having one positive charge), the effect of Na ions (increasing permeability) is counterbalanced by the effect of Ca ions (decreasing permeability) at 2 : 1 ratio of Na : Ca. Thus one Ca ion antagonises two Na ions.

N.B. The above experiment may also be performed with potato discs by weighing method. Here at the null point there will be no change in weight.

3. Experiments on osmosis

(*i*) DEMONSTRATION OF THE PHENOMENON OF OSMOSIS

EXPERIMENT: A small funnel is taken and its broad mouth is closed with a piece of parchment or egg membrane. It is then completely filled with 1 M glucose solution (180·06 gm/litre). The nose of a 10 ml pipette is fitted with the stem of the funnel with the help of a rubber tubing. The level of the sugar solution is brought to a visible mark by adding more sugar solution drop by drop through open end of the pipette. The apparatus is then placed over a beaker containing pure water and clamped properly. The increase in the level of sugar solution is noted at definite intervals.

OBSERVATION: The level of sugar solution increases in the pipette gradually. The rate of this increase declines with time. A positive test for glucose (a brick red ppt.) is obtained when an aliquot of water from the beaker is tested with Fehling solution (Mix Fehling A containing 35 gm $CuSO_4$ plus 500 ml water, and Fehling B containing 50 gm NaOH plus 173 gm Rochelle salt plus 500 ml water, in equal proportions, add the test solution to it and heat strongly).

INFERENCE: The following inferences can be drawn from this experiment:

(*a*) The sugar solution rises in the pipette because of accumulation of water molecules which pass through the semipermeable membrane due to endosmosis.

(*b*) The accumulation of water dilutes the osmotic conc. of sugar solution. Hence the rate of increase of the level of sugar solution inside the pipette decreases with time.

(*c*) The positive reaction of sugar in the water of the beaker indicates that some sugar molecules have also come out through the membrane by exosmosis as the membrane is not truly semipermeable but differentially permeable.

(*ii*) DEMONSTRATION OF OSMOSIS BY OSMOSCOPES

A. EGG OSMOSCOPE

EXPERIMENT: The inner contents of an egg are taken out through a small hole made at one end of it. To obtain the semipermeable membrane, about one-third of the shell is immersed in conc. HCl very carefully. The acid dissolves the shell (made up of $CaCO_3$) exposing the inner membrane of the egg. It is then washed well with water without damaging the semipermeable membrane. The nose of an one millilitre pipette is inserted through the hole of the shell up to some distance avoiding contact with the membrane, and sealed with sealing wax or lacre. The egg is filled with a strong solution of 1M sucrose (342·30 gm/litre) through the open end of the pipette and the level is brought to a visible mark on the pipette which is noted. Now, the egg membrane of the setup is immersed in a beaker of pure water and clamped vertically (Figure 3).

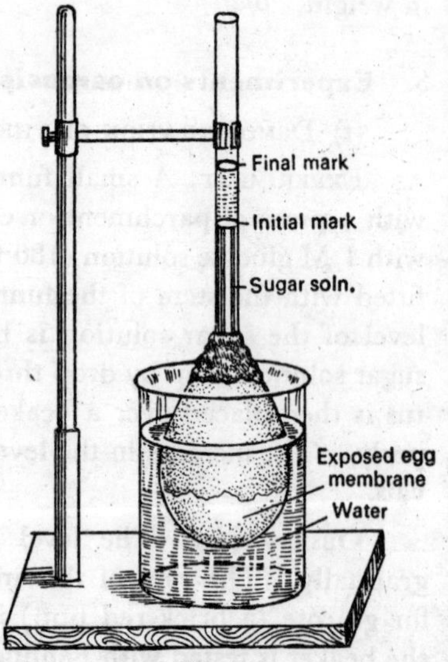

Fig. 3.

OBSERVATION: After some time the liquid inside the pipette rises. This rise declines with time.

INFERENCE: As in Expt. 3 (*i*).

N.B. When the egg osmoscope is placed in an isotonic medium, there will be no rise of liquid in the pipette; in a hypotonic medium the level will rise due to endosmosis; and in a hypertonic medium the level of liquid will fall due to exosmosis.

B. POTATO OSMOSCOPE

EXPERIMENT: A large potato tuber is first skinned and cut into a rectangular shape. A simple well is made at the centre of the tuber with the help of a cork borer and a scalpel without piercing the other side. This potato osmoscope is then half-filled with 1 M sucrose solution, its level is marked with a pin and is placed in a petridish containing pure water.

OBSERVATION: After some time the level of liquid rises in the osmoscope.

INFERENCE: As in Expt. 3 (*i*). Here the potato tissue is acting as a semipermeable membrane.

N.B. The potato osmoscope may be placed in isotonic, hypotonic and hypertonic solutions and the corresponding changes in the level of liquid in the osmoscope may be observed.

(*iii*) DEMONSTRATION OF PLASMOLYSIS AND DETERMINATION OF ISOTONIC CONC. OF THE CELL SAP

EXPERIMENT: Epidermal peelings from the lower surface of *Rhoeo* or *Tradescantia* leaves are suitably employed in this experiment because the protoplasm of the cells is clearly visible due to presence of anthocyanin pigment. *Spirogyra* filaments may also be used in this experiment.

0·15, 0·20, 0·25, 0·30, 0·35 and 0·40 M sucrose solutions are prepared from 1 M solution. 2 ml of each is poured in separate watch glasses and a few peelings are immersed in each of the solutions. After about 40–50 minutes one peeling is taken out from each solution, mounted in the same solution and examined under the microscope to study the number of cells plasmolysed under the field of vision.

OBSERVATION: Two concentrations are found out, one where plasmolysis has just started (i.e., slightly stronger than the cell sap) and the next lower concentration which just fails to plasmolyse the cells. Average of the two concentrations gives approximately the equivalent concentration of the cell sap. The extent of plasmolysis (both shrinkage of protoplasm and number of cells plasmolysed) also varies in different molar solutions.

INFERENCE: Plasmolysis of the cells does not occur in molar solutions of sucrose which are hypotonic and isotonic to the cell sap. Plasmolysis occurs only in case of hypertonic solutions. Incipient plasmolysis occurs at a concentration which is just above the isotonic concentration.

N.B. It is not possible to determine the exact concentration at which incipient plasmolysis (turgour pressure is zero) occurs. Thus the average of the concentrations mentioned above is taken as isotonic concentration. In case of individual cells, this method of determination of isotonic conc. should be adopted.

In case of integrated plant tissue, isotonic concentration may be obtained by taking the average of two concentrations of sugar solution where plasmolysis has just started and where cent percent cells plasmolysed. This concentration can be directly determined from a curve (percentage of cells plasmolysed/concentrations of sugar solutions) taking the concentration to be isotonic where 50% of the cells plasmolysed (Figure 4).

When the plasmolysed cells are put in pure water, deplasmolysis takes place that is protoplasm regains its original shape. If the concentration of the sugar solution is too strong, no deplasmolysis will take place indicating death of protoplasm.

This experiment may be performed with coloured flower petals, filamentous algae like *Spirogyra*, etc. The pattern of shrinkage of protoplasm is specific for each species of plant.

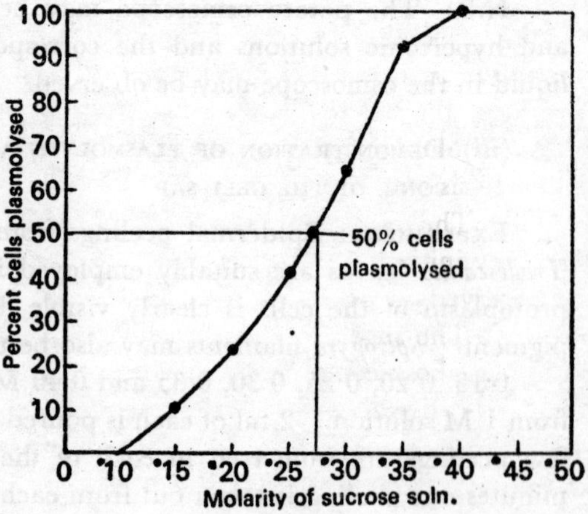

Fig. 4.

(*iv*) DETERMINATION OF OSMOTIC PRESSURE OF INTEGRATED PLANT TISSUES

EXPERIMENT: Lower epidermal peelings of *Rhoeo*, *Tradescantia* or filamentous algae like *Spirogyra* may be used for this experiment. The isotonic concentration may be determined as in Expt. 3 (*iii*).

RESULTS: Osmotic pressure of the cells is calculated by either of the following formulae:

(*a*) $$\text{O. P.} = \frac{22{\cdot}4 \times \text{M} \times \text{T°C.}}{273}$$

where O. P. = osmotic pressure in atmosphere,
M = osmotic concentration of the cell sap in molarity,
T°C. = 273 + laboratory temperature,
273 = absolute temperature,
22·4 = osmotic pressure of 1 M sugar solution at N.T.P.

(*b*) $$\text{O. P.} = \text{C} \times \text{R} \times \text{T}$$

where O. P. = osmotic pressure in atmosphere,
C = osmotic concentration of the cell sap in molarity,
R = universal gas constant (0:082),
T = 273 + laboratory temperature.

DISCUSSION: Osmotic pressure thus determined gives the approximate value since the accurate concentration of the cell sap cannot be determined using plasmolytic method due to several reasons. Some of the important sources of error are as follows:

(*a*) For normal and ordinary cells the actual isotonic conc. of the cells can only be obtained by correcting for the changes in volume of the cells associated with changes in osmotic conc. (see N.B.).

(*b*) Adhesion of protoplasm to the cell wall in certain types of cells may result in excessive osmotic pressure values. To get an accurate osmotic pressure, the calculated pressure must be multiplied by the ratio of the normal volume to the minimum volume of the cell, i.e., at the stage of plasmolysis (see N.B.).

(*c*) The exact determination of incipient plasmolysis is difficult because the very initiation of the separation of the protoplasm from its wall cannot be observed clearly.

(*d*) The sugar solutions may change the permeability of the cell wall bringing about considerable errors.

(*e*) Cell wall may be impermeable to the plasmolysing external solutions and may get twisted.

(*f*) Abnormal conditions of the protoplasm may result from mechanical shock due to cutting or isolation of tissues.

(*g*) The intact cells may be adversely affected by the sap released from the cut cells.

(*h*) Elastic nature of the cell wall may lead to inaccurate results.

N.B. The volume correction is made as follows:

$$C = C_i \ (V_i/V)$$

where C = conc. at normal condition,
C_i = conc. where plasmolysis has just started,
V = volume at normal condition (integrated cells),
V_i = volume at conc. where plasmolysis has just started.

Since measurement of volume is impossible due to irregular shape of the cells, the average area (length × breadth) may be taken into consideration.

Osmotic pressure may also be determined by ***cryoscopic method***. The following equation relating to freezing point depression and osmotic pressure may be employed:

$$\text{O. P.} = \frac{22 \cdot 4 \times A}{1 \cdot 86}$$

where A is the freezing point depression. A is determined as follows:

1 molar solution of a nonelectrolyte freezes at a temperature of $-1 \cdot 86°$C. and its theoretical O. P. is 22·4 atmospheres. For actual determination of A of cell sap, the cell sap at first must be extracted by killing the cells by freezing. The vessel containing the experimental tissue is surrounded with a mixture of ice and common salt. The frozen tissue is then allowed

to melt. The sap can now be easily extracted with mere hand press. The freezing point of the macerated material may be determined.

Molecular weight of a substance may be calculated from the following formula, when strength of the solution in per cent or gm/litre, osmotic pressure and room temperature are known.

$$PV = (g/M) \times RT$$

$$\text{or,} \quad M = \frac{g R T}{P V}$$

where g = gm of substance,
R = universal gas constant (0·082),
T = 273 + laboratory temperature,
P = osmotic pressure,
V = volume in litre,
M = molecular weight of the substance.

Integrated tissue like potato tuber may also be taken for this experiment of osmotic pressure determination. Here, isotonic concentration can be obtained by taking the concentration of sugar as isotonic where no change in weight takes place.

(*v*) Determination of mean suction pressure or diffusion pressure deficit of plant cells

Experiment: 20 ml each of 0·15, 0·20, 0·25, 0·30, 0·35 and 0·40 molar concentrations are prepared from 1 M stock solution of sucrose. All suberized peripheral layers from a fresh potato tuber or beet root are removed and cut into 1 cm × 2 cm pieces using a cork borer and a scalpel. Each piece is washed in distilled water, blotted well, weighed (initial weight) and transferred to each of the graded solutions of sucrose. These are then left for about an hour. After the stipulated period the pieces are taken out, blotted carefully and reweighed (final weight). In the above way, the osmotic concentration of fully turgid potato tuber tissue (made turgid by keeping the tissue in water for half an hour previously) and that of partially dehydrated tissue (made partially dehydrated by keeping the tissue in open air for half an hour) are separately determined. The differences between the osmotic pressures obtained in case of dehydrated and normal tissues and that between normal and turgid tissues give the D.P.D. of dehydrated and normal tissues respectively.

Results: The percentage increase or decrease in weight of the tissue is determined for each concentration of sucrose solution. Results are plotted on a graph paper taking sugar concentrations as abscissa and the percentage change in weight as ordinate. The concentration of sucrose corresponding to no-change-in-weight is found out from the graph. This value gives the osmotic concentration of the cell sap when osmotic pressure is equal to the suction pressure (D.P.D.), turgour pressure being zero. The D.P.D. is calculated by applying the formula of osmotic pressure given in Expt. 3 (*iv*). The necessary temperature and volume corrections may also be made.

DISCUSSION: The D.P.D. is obtained here indirectly because the D.P.D. of a solution is equal to its osmotic pressure at the isotonic concentration of the cell sap, i.e., at an equilibrium point. This value is obtained from the molar strength of the sucrose solution with which cells or tissues come to equilibrium at incipient plasmolysis, and when the cell volume is at its relative minimum due to elimination of turgour pressure brought about by exosmosis of water. The suction pressure values may be identical with osmotic pressure values only if the cells are initially at the incipient plasmolysis.

(*vi*) DETERMINATION OF TURGOUR PRESSURE OF PLANT TISSUE

EXPERIMENT: The isotonic concentration of a suitable plant tissue is determined according to Expt. 3 (*iii*) or 3 (*v*), whichever is convenient. The osmotic pressure of the tissue (which is here equal to the suction pressure or D.P.D., the turgour pressure being zero) is first determined. This is taken as the osmotic pressure (P) of the cell sap at that stage. The tissue is then kept in pure water to allow it to increase in turgour pressure. After 10 minutes the tissue is taken out and osmotic pressure (suction pressure or D.P.D., turgour being zero) is similarly determined. The process is repeated until the suction pressure reaches a constant value and corresponding suction pressures are recorded.

RESULTS: Turgour pressure is obtained by the formula, $T = P - S$ where P = osmotic pressure of the cell sap (taken as constant), S = suction pressures at different time intervals, and T = turgour pressure at a particular time. Thus difference between the initial osmotic pressure (P) and that determined at every 10 minutes interval (S) gives the rate of increase in turgour pressure (T).

DISCUSSION: When a solution is confined within a semipermeable membrane immersed in water, there will be a net movement of water molecules through the membrane, because the diffusion pressure of water molecules is greater on water outside than that of the solution inside the membrane. The inward passage of water thus develops turgour pressure within the cell. As the turgour pressure increases, suction pressure (D.P.D.) falls. In the present experiment the initial osmotic pressure of the cell has been taken as its maximum osmotic pressure and at isotonic concentration of the external solution this equals the D.P.D. of the cell (turgour being zero). When immersed in pure water again, the turgour pressure increases due to endosmosis and as a result the suction pressure falls. The osmotic pressure determined at this stage using isotonic concentration (which must be less than the initial value) gives the suction pressure. This value subtracted from the initial osmotic pressure value, gives turgour pressure at that particular suction pressure of the tissue. When the maximum turgour pressure of the cell has reached, no further change in suction pressure takes place.

4. Experiments on imbibition

(*i*) DEMONSTRATION OF IMBIBITION PROCESS

EXPERIMENT: A cubical piece of a potato tuber and that of a wood are taken, their volumes (by displacement of water) and weights are noted. These are immersed in a bottle of water (if the piece of wood floats, a sinker may be used). After 2 hours the pieces are taken out, blotted well and the final volumes and weights of each are recorded.

RESULTS: The difference between the final and initial volumes and that between weights for each type of material gives the amount of water imbibed by them.

DISCUSSION: Imbibition is a physical process of diffusion and is independent of the presence of a differentially permeable membrane which is the characteristic of all living cells. The amount of liquid imbibed depends upon: (a) the diffusion pressure gradient between the imbibant and the liquid imbibed, and (b) the affinity between the components of the imbibant and the liquid. Hence, there may be a difference in the initial rate of imbibition between a piece of wood and a potato tuber when imbibing medium is pure water because in the former case only imbibition is operative while in the latter case both imbibition and absorption by osmosis are in action.

N.B. Imbibition is influenced by some physical conditions of the external medium such as pH, temperature, etc. The osmotic concentration of the imbibing medium also affects the rate of imbibition because this rate depends upon the diffusion pressure gradient between the imbibant and the imbibing liquid. Imbibition differs from osmosis in that, when the colloid absorbs liquid and swells, considerable solution may be absorbed, and not just water as in an osmotic system having a semipermeable membrane. (For example, blocks of gelatine or agar may swell nearly as much in a half-molar solution of sucrose as in pure water).

Experiments may be suitably designed to study the effects of these external factors on the rate of imbibition by different types of plant materials.

(*ii*) DEMONSTRATION OF THE EFFECT OF OSMOTIC CONCENTRATION (OR OSMOTIC PRESSURE) ON IMBIBITION

EXPERIMENT: Graded molar solutions of NaCl within the range of 0·1 to 1·0 M are prepared. Several cubical pieces of potato tuber and wood are prepared. Each of such pieces is initially weighed or volume measured and immersed individually in petridishes containing different graded solutions. Water control for each type of material is also kept. Final weights or volumes of the pieces for each concentration are taken after 2-3 hours and the difference in final and initial weights or volumes in each case is calculated. A graph is plotted taking concentration as abscissa and percentage increase (or decrease) in weight or volume as ordinate.

RESULTS: From the curve it is observed that there is increase in percentage weight or volume in case of both potato tuber and wood but the amount of water imbibed decreases with increasing molarity of NaCl. In case of potato tuber (a living tissue), a point (isotonic conc.) is reached where no change in weight or volume takes place and beyond this point (hypertonic conc.) there is a decrease in percentage weight or volume but in case of a piece of wood there is no such effect.

DISCUSSION: In case of a piece of wood (nonliving) water enters by imbibition only which is dependent upon the diffusion pressure of the imbibing medium. The addition of a solute to pure water lowers its diffusion pressure. The decrease in diffusion pressure gradient decreases the rate at which water is imbibed and the amount of water taken up. But in case of potato tuber consisting of living cells, having semipermeable membrane, diffusion of water is mainly controlled by osmosis. Hence diffusion of water takes place due to endosmosis so long the external solution remains hypotonic but as soon as an isotonic concentration is reached no change in weight or volume takes place. At a hypertonic concentration, weight or volume decreases due to exosmosis.

(*iii*) DEMONSTRATION OF PRESSURE RELEASED DUE TO IMBIBITION

A. IMBIBITION PRESSURE

EXPERIMENT: A suitable can with a lid (cigarette tin) may be suitably employed for this experiment. A few small holes are made at its bottom and is completely filled with dry gram seeds. The lid is then fixed tightly and the bottom of the can is then placed in a petridish containing water.

RESULTS: The lid comes off forcefully by itself after some time and the time required for this is noted.

DISCUSSION: The dry seeds swell on coming in contact with water below. The swelling results in development of considerable imbibition pressure which pushes the lid to come off.

N.B. Some known weight may be placed on the top of the lid an imbibition pressure required to just overcome this may be taken as a measur of imbibition pressure.

B. MEASUREMENT OF IMBIBITION PRESSURE

EXPERIMENT: Some dry seeds are taken in a glass jar and a few ml o water are poured into them. The seeds are then covered with a flat disc which is centrally fitted with a graduated piston. The piston passes through a cork fitted at the mouth of the jar. Care should be taken that the graduated piston moves easily through the cork. The top of the piston is fitted with a small disc so that different weights can be placed on it. The initial mark of the piston just above the cork is noted (Figure 5).

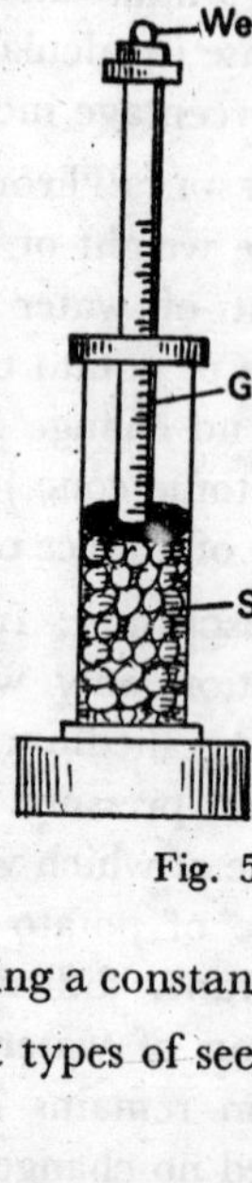

Fig. 5.

RESULTS: As the seeds swell due to imbibition after some time, the graduated piston moves upward and gives the indication of imbibition pressure. Now weights are placed on the top of the disc to bring the piston to its initial mark. This amount of weight gives a measure of the imbibition pressure. The pressure is determined at different time intervals using different weights required to bring the piston down to the initial mark.

DISCUSSION: The swollen seeds exert imbibition pressure on the disc which pushes the piston upward indicating imbibition force. The rate of imbibition pressure gradually increases with time ultimately reaching a constant value.

N.B. The rate of imbibition pressure for different types of seeds may be compared with the help of this experiment.

(*iv*) DEMONSTRATION OF RELEASE OF HEAT ENERGY DURING IMBIBITION

EXPERIMENT: 200 gm of starch is weighed out quickly which was previously dried in an oven at 100°C., transferred to a desiccator and cooled to room temperature which is recorded. The starch is placed in a clean dry thermosflask and 250 ml of water is immediately added. It is then thoroughly mixed with the help of a glass rod and the mouth of the flask is closed tightly with a cork through which passes a thermometer and its initial reading is recorded. The rise in temperature by the starch water mixture is recorded after a few hours of imbibition.

DISCUSSION: The imbibed water is actually compressed occupying less space than when present as free water. Part of the water is adsorbed with great force and thus concentrated at the colloidal interfaces. This adsorption releases the potential energy of the system which was present before the water was placed in contact with the colloids. The large amount of heat evolved when water is added to such colloidal materials which have been first thoroughly dried offers a demonstration of this change from potential to kinetic energy in the form of heat. Thus the surface energy within the colloid resulting from withdrawal of water and drying is released when water is again added.

(v) Demonstration of imbibition of water by different types of seeds

Experiment: This experiment may be performed with three types of seeds, viz., proteinaceous pea, starchy rice and fatty groundnut. Seed coats are removed and 4 ml of each type of seeds is taken (by displacement of water in a measuring cylinder the volume may be measured). These are then blotted well and weighed. Each type of seed is then taken in a beaker and equal volume of water is poured in each so that the seeds are completely immersed. The seeds are then taken out at an interval of 15 minutes, blotted properly and increase in weight is noted.

Results: The percentage increase in weight by each type of seed is determined and plotted graphically.

Discussion: When dry seeds are put in water after removing their coats, which render a physical barrier to absorption, water is absorbed by the seeds mainly by imbibition owing to the colloidal nature of the protoplasm of seeds. As a result seeds become turgid and osmotically active and absorption of water is then regulated by osmotic process. The rate and amount of absorption by different types of seeds vary because of their different protoplasmic constituents which control their imbibing capacity. Results of the present experiment show that the percentage of absorption of water is maximum in case of proteinacous seeds and minimum in case of fatty seeds. Proteins and polypeptides are hydrophilic colloids which have strong affinity for water. Cellulose and starch also absorb water strongly to the surfaces of these hydrophilic colloids.

N.B. (*a*) Imbibition by intact seeds leads to considerable error due to formation of water layer in the air space between the endosperm and the seed coat. In order to determine the imbibition of water by intact seeds amount of imbibition without seed coat is to be added to the amount of imbibition by the seed coat only.

(*b*) Effect of seed coat on imbibition can be studied by setting experiment with seeds with and without seed-coats.

(*c*) Q_{10} of absorption of water can be studied by allowing the seeds to imbibe water at room temperature, 10°C below and above it (see Expt. 3 (*a*) of Chapter XI).

CHAPTER III

LIVING NATURE OF PROTOPLASM

There are two main groups of objects in the world, one group includes 'the living' and the other 'the nonliving'. A correct definition of a living matter is not yet possible, because this involves a complete knowledge and understanding of it. Nevertheless, since we are able to distinguish between 'animal, vegetable and mineral', it is logical to ask upon what criteria such judgement rests. We usually answer this question by listing a number of characteristics that we ascribe to living matter, such as movements, reproduction, metabolism, sensitivity, growth, etc. It is generally admitted that these properties are not exhibited in the nonliving world. But it is argued that in living matter they are present simultaneously and find their most complex degree of expression. So far the most consistent and logical definition of life has been given by John Perret which reads as 'life is a potentially self-perpetuating open system of linked organic reactions, catalysed stepwise and almost isothermally by complex and specific organic catalysts which are themselves produced by the system' and is keeping with the phenomenon and characteristics of living organisms.

All living matters are composed of cells and each cell is the biological unit of activity delimited by a semipermeable membrane and capable of self-reproduction. Though it is divisible into subcellular particles that retain some of the properties of original functional unit, the cell nonetheless represents a minimum unit of independent activity. Protoplasm is the name given to the living material present in the cell. Protoplasm is enclosed within the boundary of the cell wall and fluidity is its common state. Properties of life in the protoplasm are governed by the physical and chemical laws without which the maintenance of life would have been difficult and doubtful.

In this chapter some experiments are selected to show some of the characteristics of living nature of protoplasm, and one of such characteristics is its responsiveness towards some external physical and chemical stimuli.

1. **Demonstration of living nature of protoplasmic membrane**

(*i*) EFFECT OF TEMPERATURE ON PERMEABILITY

(*a*) Perform Expt. 2(*i*) of Chapter II.

(*b*) EXPERIMENT: A few *Spirogyra* filaments or *Rhoeo* peelings are mounted in water on a slide and one end of it is heated carefully over a flame. It is then examined under a microscope and any change in the appearance of the cell is noted. The filaments are dipped in water and examined again under the microscope.

OBSERVATION: Shrinkage of protoplasm (plasmolysis) takes place. In case the cells are dead, no deplasmolysis takes place when dipped in water.

INFERENCE: Drastic plasmolysis takes place due to destruction of the living nature of plasmamembrane by high temperature treatment. There is no deplasmolysis and the cells do not revive when placed in water showing irreversible changes in the protoplasm and destruction of plasmamembrane brought about by high temperature.

(*ii*) EFFECT OF HIGH CONCENTRATION OF SOLUTES ON PERMEABILITY

EXPERIMENT: Some *Spirogyra* filaments or *Rhoeo* peelings are kept in 2 M sucrose solution for about 15 minutes and then examined under the microscope. The tissue is then taken out and placed in distilled water for some time. This is again observed under the microscope.

OBSERVATION: In a strong solution of sucrose cells show pronounced plasmolysis. When kept in water cells do not revive, i.e., there is no deplasmolysis.

INFERENCE: The strong solution of sucrose causes irreversible injury to the plasmamembrane and the protoplasm of cells leading to their destruction.

(*iii*) EFFECT OF TOXIC SUBSTANCES ON PERMEABILITY

EXPERIMENT: Four skinned slices of beet root are washed thoroughly with water and dipped in four test tubes containing (a) 5% solution of H_2SO_4, (b) 5% solution of $CuSO_4$, (c) 5% ethanol, and (d) distilled water and kept for about 40 minutes.

OBSERVATION: The red anthocyanin pigment diffuses out in all the cases excepting in the tube containing distilled water.

INFERENCE: The semipermeable nature of the plasmamembrane is lost due to treatments with different toxic substances and thus the colour diffuses out. In case of distilled water semipermeable nature is not affected and the colour does not come out.

2. Experiment on protoplasmic streaming

EXPERIMENT: A portion of the peeling of freshly collected *Vallisneria* is mounted on a slide and is slightly warmed. It is then examined under the high power of microscope.

OBSERVATION: The rotatory movement of the chloroplastids along the boundary wall of the cell is observed.

INFERENCE: The protoplasm exhibits streaming movements particularly in plant cells with large vacuoles indicating its living nature. The exact cause of this movement is not yet clear. The exogenous supply of energy, however, enhances this rate and an uncoupler of oxidative phosphorylation like DNP (Dinitrophenol, 184·11 mg/litre) decreases this rate.

N.B. Experiments can be suitably performed using the cells of stamen hair of *Zebrina* or *Rhoeo* and cells in the young tips of *Chara* or *Nitella* and the root cells of *Limnobium spongia*.

3. Demonstration of living nature of protoplasm by vital staining

EXPERIMENT: Some epidermal peelings of onion scales are immersed in a few ml of a 0·005% neutral red solution (dissolve 0·005 gm neutral red in 100 ml 50% ethanol) in a watch glass. After 1, 5, 10 and 20 minutes interval peelings are taken out from the dye, rinsed in distilled water, mounted on slides and examined under the high power of microscope.

OBSERVATION: Protoplasm of the cells are coloured red and the intensity of colour deepens with increase in time of treatment.

INFERENCE: Neutral red is a vital stain and all living protoplasm is stained by it. Prolonged treatment with the dye causes more adsorption of it on the protoplasm and the colour is intensified.

4. Experiment to show the phenomenon of respiration in plants.

Refer Chapter X.

5. Experiment to show the phenomenon of growth in plants.

Refer Chapter XII.

6. Experiment to show the phenomenon of movement in plants.

Refer Chapter XIV.

7. Experiment to show the phenomenon of reproduction in plants.

Refer Chapter XV.

CHAPTER IV

SOIL AND WATER RELATIONS OF PLANTS

The land plants depend upon the source of water present in the soil, which is absorbed by the roots. The root system usually comprises not more than a quarter of the whole plant on dry weight basis, but they are so finely divided that they may occupy a mass of soil greater than the volume of atmosphere surrounding the shoot. Thus there is a tremendous amount of surface contact between the soil and the plant. Land plants are so completely dependent upon the soil for anchorage, water and nutrients that the factors of the soil are of supreme importance not only due to this intimacy of contact between the plants and the soil, which are strongly influenced by each other, but also due to greater complexity of the structure of the soil. Therefore, the nature and structure of the soil and form of existence of water in them are important aspects of study of soil and water relations of plants.

Soil contains both organic and inorganic substances in solid, liquid or gaseous states. Small and big rock particles, organs of living or dead plant materials, microorganisms, water and air all are present in the soil. These various components are mixed in different relative proportions in different types of soils. The physical and chemical components of soils produce direct or indirect effects on plants growing in a particular soil. The texture of a soil depends primarily on the sizes of its mineral particles and this feature is of considerable importance to vegetation because the texture of the soil controls aeration and water holding capacity of a particular soil. On the basis of the size of particles, soil is separated into coarse soil (particles above 5 mm diameter), gravel (5–2 mm diameter), coarse sand (2–0·2 mm diameter), fine sand (0·2–0·02 mm diameter), silt (0·02–0·002 mm diameter) and clay (below 0·002 mm). A loam is a soil which consists of a thorough mixture of particles of widely different sizes with adequate humus fraction.

Water holding capacity of soil depends upon the texture, i.e., on the size of the soil particles and the amount of humus present. The coarser the particles are, the larger is the air spaces and less is the water holding capacity and conversely. The amount of water in a soil depends upon environmental conditions, area and kind of soil. The major portion of rain water flows away along the slopes of inclined soil and this is called *runaway water* which becomes unavailable to plant root. Some rain water percolates through the soil by the force of gravity and reaches the water table below and this is called *gravitational water* which is also unavailable to plant roots. But during percolation some water is retained in the soil as *capillary water* and as *hygroscopic water* against the force of gravity. Capillary water fills the capillary spaces between the soil particles by cohesive forces and this is available to plant roots. Hygroscopic water forms thin layers or films on the surface of hydrophilic colloidal particles of the soil by adhesive forces and unavailable to plant roots. Some water may be present in a chemically bound state in the structure of soil particles and is called *combined* or *bound water* which is also unavailable to plant roots.

The amount of water in a soil which remains even against the gravitational pull is called *field capacity of the soil* and expressed as percentage of dry weight of the soil sample. The maximum *water holding capacity* of a soil is the amount of water held when the soil is saturated and the excess water is allowed to drain away by gravity. *The moisture equivalent* is defined as the amount of water held in a layer of soil 1 cm thick when saturated and subjected to a centrifugal force of 1000 × g. *The wilting coefficient* or *the permanent wilting percentage* is a physiological measure of the water relations of the soil and is defined as the percentage water content (expressed as percent dry weight) of a soil when the plant or plants growing in it has reached the conditions of permanent wilting.

The following experiments have been designed to demonstrate some of the important characteristics of soil and water relations.

1. Determination of water content of different types of soils

EXPERIMENT: 50 gm sample of each of clayey, loamy and sandy soils is taken. Soils are dried in an oven at a temperature of 105°C. for approximately 48 hours and the final weights are recorded.

RESULTS: The loss in weight determines the amount of water which was present in the soil. The percentage of water content is calculated from the following formula:

$$\text{Percentage of water content} = \frac{\text{loss in weight} \times 100}{\text{dry weight of the soil}}.$$

Loss in weight is determined by subtracting the final weight of the soil after drying from the initial weight of the soil before drying. The dry weight of the soil is determined by subtracting the weight of the container from the total weight of container plus soil after drying.

DISCUSSION: Size of the particles present in a soil determines the water holding capacity of the soil. The larger the size of the particles and their proportion, the more is its air holding capacity and the less is the water holding capacity. The sandy soil particles being of larger dimensions have poor water content. The reverse is true in case of clayey soil, and loamy soil falls in-between.

2. Determination of field capacity (or water holding capacity) of the soil

EXPERIMENT: Some sandy and clayey soils are completely dried in a hot air oven. Two long cylindrical tubes of 3 cm diameter are taken and filled with weighed amount (100 gm) of the soil samples after plugging one end with cotton. Both the tubes are clamped in a vertical position and from the open end of each water is slowly poured and allowed to percolate through the soil by gravitational pull. When the soil is saturated, pouring of water is stopped and the tubes are allowed to stand for some time till all the gravitational water has drained off. The wet soil is then taken out from the tubes and the amount of water retained by each is found out by weighing and drying method.

RESULTS: The amount of water retained by 100 gm weight of original dry soil gives the field capacity of the soil which can be calculated from the following formula:

Field capacity or percentage of water holding capacity

$$= \frac{\text{gain in weight at saturation point}}{\text{dry weight of the soil}} \times 100$$

DISCUSSION: Field capacity is a measure of the amount of water present in a given soil when it is saturated. The field capacity or water holding capacity generally depends on the nature of the soil particles, porosity, temperature and the presence of hydrophilic colloidal materials in the soil. Sandy soil has the least, clayey the maximum and loamy has the intermediate relative values of field capacity.

N.B. In this experiment some errors may creep in due to adherence of soil particles to the container as well as to the cotton plugs. This may be avoided by subtracting the weight of the cylinder plus cotton plug (previously determined) from the final weight.

3. Determination of porosity, water holding capacity and adsorptivity of soil

EXPERIMENT: 100 gm each of dry sand, dry silt (from river bed) and dry clay is taken. Three funnels having folded filter papers are separately filled with three types of soils. 100 ml of distilled water coloured with safranine are slowly poured on each type of soil and allowed to drain out through the stem of the funnel. The drainage is collected in beakers and their volumes are measured (Figure 6).

OBSERVATION: It is seen that the collected volume of water is more in case of sandy, less in silty and least in case of clayey soils. The filtered water shows decrease in the intensity of colour up to different degrees—

maximum in sandy and minimum in clayey soil. The clay soil adsorbs maximum safranine colour and sand minimum.

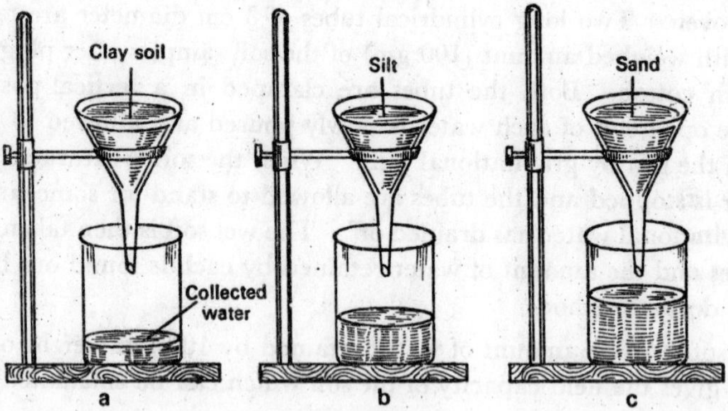

Fig. 6.

INFERENCE: The experiment shows that clay has maximum water holding capacity and adsorptivity but least porosity. Sand has highest porosity but least water holding capacity and adsorptivity. Silt shows intermediate values between sand and clay.

4. Determination of capillary flow in soil

EXPERIMENT: Three cylindrical glass tubes of the dimension 40 cm long and 2 cm diameter are taken and open end of each tube is plugged with cotton. Air dried soil samples (clay, silt and sand) are taken in each tube separately and uniformly packed up to certain height by moderate tapping. Tubes are allowed to stand vertically in a beaker containing water so that the plugged end dips in water.

OBSERVATION: Water gradually rises up through the soil in the tubes. It diffuses upward faster through sand followed by silt and clay.

INFERENCE: The capillary spaces between the colloidal particles of clay being narrower and finer in dimension exert greater resistance to flow of water and maximum adhesive force. In case of sand the capillary spaces are wider offering lesser resistance to flow, and minimum adhesive force.

5. Determination of permanent wilting percentage (PWP) or wilting coefficient of a soil

EXPERIMENT: Physical conditions of the environment should be uniform during this experiment.

Experimental soil sample is taken in a water-proof pot and a suitable plant is grown in it till it attains maturity. The soil of the pot is watered properly. The soil surface is covered with a polythene sheet to check evaporation from its surface. The only loss of water from the soil now occurs due to absorption and transpiration by the plant. Further watering of the pot is strictly avoided. As time elapses, older leaves start wilting and reach a point of permanent wilting and the plant does not recover even after being kept overnight in a chamber having 100% relative humidity.

RESULTS: The sample of soil from the pot is taken and its water content is determined by oven-drying method. The permanent wilting percentage or wilting coefficient is calculated as follows:

$$\text{PWP} = \frac{\text{amount of water in X gm of pot soil at permanent wilting point}}{\text{dry weight of X gm of pot soil at the same time}} \times 100$$

DISCUSSION: Wilting coefficient or permanent wilting percentage (PWP) indicates the amount of water in the soil which is actually unavailable to the plant when permanent wilting has set in. Different soils have different wilting coefficients and therefore affect growth of the plant to various extent. The same soil can, however, have different PWP values for different species of plants. This usually lies between 1 to 15% and depends upon the nature and texture of the soil. This includes hygroscopic water and bound water of the soil. Sandy soil has lowest and clayey soil has highest PWP values. Wilting coefficient indicates that amount of water content in a soil below which plants cannot grow and develop normally.

N.B. Wilting coefficient may also be determined from moisture equivalent of the soil which is the amount of water held in a layer of soil 1 cm thick when saturated and subjected to a centrifugal force of 1000×g. Wilting coefficient is determined by the following formula:

$$\text{PWP} = \frac{\text{moisture equivalent}}{1{\cdot}84 \pm 0{\cdot}013}$$

CHAPTER V

ABSORPTION AND TRANSLOCATION OF WATER OR ASCENT OF SAP

The process of water absorption by plants and its translocation by the conducting elements are one of the important aspects for the continual maintenance of life processes. Water is first absorbed from the soil by root hairs and other epidermal cells in and around the root hair zone. Water then moves through the cortex tissue across the endodermis and pericycle and finally into the xylem ducts. The xylem tissue of the roots connects directly the xylem tissue of the stem, thus allowing water to move out of the root and into the stem. The xylem of the stem is divided and subdivided many times to form a complex net-work of water conducting tissue in the leaf finally ending in the fine veins. Water then moves from leaf veins into the mesophyll cells, it is evaporated from their surfaces and finally goes out as water vapour through stomata into the surrounding atmosphere.

The process of water absorption is dependent directly or indirectly on the transpiration pull. The absorption may be *active* or *passive*. Active absorption involves expenditure of energy and absorption of water takes place either osmotically due to active accumulation of salts at the expenditure of energy of living cells or by root-pressure where also living cells actively participate and water is absorbed by pumping action. The passive absorption, on the other hand, is due to either transpiration pull or by cell to cell osmosis along the concentration gradient.

The absorption of water is affected by the factors like temperature of the soil, concentration of the soil solution, aeration of the soil, availability of soil water and factors influencing transpiration.

The experiments incorporated in the following pages will demonstrate the phenomenon of water absorption by plants, its regulation and the factors influencing the process of absorption.

1. Demonstration of the phenomenon of passive absorption in plants.

(*i*) ABSORPTION OF WATER BY PLANTS

EXPERIMENT: A *Peperomia* or a tube-rose with its root system is inserted through a cork in a conical flask containing 2% solution of eosin. The whole set-up is kept in an open atmosphere for some time. A second similar set-up is also kept under a humid bell jar.

OBSERVATION: After about an hour it is observed that considerable translocation of coloured water takes place in the first set-up whereas in case of the second set-up it is significantly inhibited. In the experiment with tube-rose white flower petals turn pink sooner in the first than in the second set-up.

INFERENCE: The experiment indicates clearly that the red-coloured eosin solution has been absorbed by the root system and translocated upward through stems and leaves. This is a case of passive absorption which depends upon transpiration pull. A transverse section of the stem shows that the lumen of the xylem vessels is coloured red. In the second set-up which was kept in humid condition, the transpiration is considerably checked and hence the absorption and translocation are delayed.

(*ii*) RATE OF WATER ABSORPTION BY PLANTS

EXPERIMENT: A wide-mouthed graduated bottle fitted with a cork is taken. It is filled with water and a rooted plant is inserted through the central hole of the cork so that the roots dip in water. All connections are made air-tight. The initial water level is marked and the whole set-up is kept in an open atmosphere for a few hours. The level of water is noted at an interval of half an hour.

OBSERVATION: The level of water in the bottle gradually decreases with time.

INFERENCE: Water is absorbed by the roots, translocated through the stem and ultimately reaches the leaves from where it is lost due to transpiration. A suction force is thus developed and the absorption is maintained. If the factors influencing transpiration are not disturbed, the rate of absorption is uniform throughout. The level of water in the bottle decreases due to loss by transpiration.

N.B. The influence of different plant parts and other factors on the rate of absorption of water may be studied as follows:

(*a*) Effect of number of leaves.
(*b*) Effect of dry and humid atmosphere by keeping the plant in dry or humid condition.
(*c*) Effect of dissolved solutes in water.
(*d*) Effect of absence of air in water.
(*e*) Effect of cutting off the roots.
(*f*) Effect of removal of root hairs.

2. Demonstration of the phenomenon of active absorption in plants

EXPERIMENT: A healthy potted tomato or *Vinca* plant is selected and the pot is watered well before the experiment is started. The plant is detopped 4 cm above the soil surface. The stump of the decapitated plant is tightly fitted with a graduated glass tube with the help of a rubber tubing. The tube is one-fourth filled with water and its initial level is noted. After 1 or 2 hours the rise in water level in the tube is again noted (Figure 7).

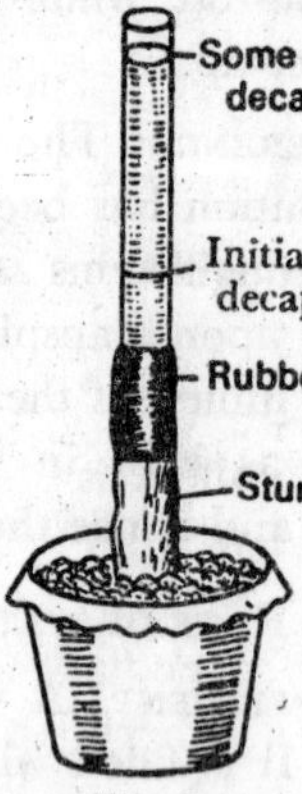

Fig. 7.

OBSERVATION: It is observed that the water in the tube is pushed up after some time.

INFERENCE AND DISCUSSION: The rise of water in the tube is due to root pressure which is responsible for the upward movement of water through the xylem vessels. The difference between the initial and final readings of the level of water gives the quantitative measurement of the root pressure exerted by the detopped plant within that period.

Root pressure is a pressure which develops in the tracheary elements of the xylem as a result of the metabolic activity of roots. It is, therefore, referred to as active process. The movement of water up the stem as a result of root pressure is due to osmotic mechanisms (passive), which are created as a consequence of active absorption of salts by the roots. Root pressure is thus referred to as being an active process in the sense that living roots are essential for it. Root pressure appears only when the plant is transpiring at slower rate but absorbing water at a higher rate. In highly transpiring plants, a negative root pressure develops.

N.B. That the absorption of water takes place actually at the expense of respiratory energy can be shown by inhibiting the respiration of roots by treatments with poisons like potassium cyanide (KCN), chloroform ($CHCl_3$), ether [$(C_2H_5)_2O$] etc.

3. The role of root growth in absorption of water

EXPERIMENT: A glass cylinder (5 × 12 cm) is uniformly filled with air-dry soil. Several seeds of *Phaseolus* or *Vicia* are sown 2 cm below the soil surface. Sufficient water is added to moisten the soil. When the hypocotyls come out, the soil surface is covered with a polythene sheet to check surface evaporation. The polythene sheet contains as many holes as there are hypocotyls so that they may pass through these holes. The holes should be plugged with raw cotton after the emergence of the hypocotyls. five seedlings are taken out at a time, washed well to remove adhering soil particles, length and number of roots are measured and fresh and dry weights (at 100°C. for 48 h) are taken. This is repeated every alternate day for 8 days. Water content is determined from the difference of fresh and dry weights.

OBSERVATION: The water content is found to increase with increase in growth of the roots.

INFERENCE: The absorption of water is found to be directly correlated with the growth of the root system.

N.B. This experiment may also be performed in the following way: Sixteen uniformly growing seedlings of the same age are taken out from the pots and washed well to remove the adhering soil particles. Four conical flasks are filled with water and weighed individually. Four intact seedlings are put in one flask so that the roots are completely immersed. Half of the roots of another lot of four seedlings is removed and put in the second flask. Similarly three-fourth of the root portion is removed from another lot of four seedlings and put in the third flask. Roots are completely removed from the base of the remaining lot of four seedlings and placed in the fourth flask. After two hours all the seedlings are taken off carefully and each flask is reweighed. The difference between the initial and the final weights in each case indicates the amount of water absorbed by each lot of seedlings. The results may be expressed as the amount of water absorbed in gm per gm fresh weight of the seedlings in each treatment. It will be observed that the quantity of water absorbed by the seedlings depends upon the amount of roots present.

4. Influence of water content of the soil on the rate of absorption of water

EXPERIMENT: Three suitable potted plants of similar age and growth are selected so that each plant bears at least a few fully grown leaves. Older leaves which appear chlorotic may be removed. The soil of the first pot should be saturated with water, that of the second pot moderately watered (preferably a little above field capacity). The third pot should remain unwatered (preferably a little above wilting percentage). Arrangements should be made to check surface evaporation from the soil. Pots are kept under bright sunlight and the experiment is started preferably in the morning. Weighings of the pots are made every two hours interval for a period of at least 8 hours. At the end of the experiment aggregate leaf area of each plant is determined (Refer Expt. 5 of Chapter VI). The rate of transpiration per square centimeter leaf area for each 2 hour interval is calculated. A graph is plotted taking the time as abscissa and amount of transpiration per sq. cm as the ordinate.

OBSERVATION: The rate of transpiration is maximum in the first pot, intermediate in the second and minimum in the third.

INFERENCE: In this experiment it has been assumed that the rate of transpiration is a measure of the rate of absorption. In case of the first pot, maximum absorption of water takes place because maximum available water is present. In the second pot having moderate water content shows intermediate absorption while the third pot being relatively dry shows minimum absorption of water.

5. Influence of temperature of the substratum on the rate of absorption of water

EXPERIMENT: Three suitable wide-mouthed bottles are filled with water. Water of the first bottle is kept at a temperature of 10°C, the second at about 25°C. or room temperature and the third at about 40°C. The temperature may be maintained using ice and hot water. Three similar plants of any herbaceous species (*Coleus*, *Phaseolus* or *Helianthus*) with large, easily measurable leaves are selected. Plants which have been grown in water culture (see Chapter VII) should preferably be used in this experiment but pot-grown plants may also be used if great care is taken to avoid damage of the root system. Roots of one plant are inserted in each bottle so that it is completely immersed in water. Each plant is supported in its respective bottle with cotton plug. The cotton plug is covered with polythene sheet in order to check evaporation from the bottle. Each bottle is weighed and kept under bright sun light. The bottles are reweighed separately at an interval of 1 hour for a period of 4 hours. The rate of transpiration per sq. cm of leaf area per hour is calculated. A graph is plotted in a similar way as done in the previous experiment.

OBSERVATION: The rate of absorption increases with rise in temperature of the substratum.

INFERENCE: Here the rate of transpiration has been taken as an index of rate of absorption. When substratum temperature decreases, rate of water absorption by roots is also decreased. Changes in substratum temperature affect the growth rate of roots, movement of water in the soil, permeability of membranes and walls of root cells, viscosity of protoplasm of root cells, viscosity and density of water, vapour pressure of water, rate of metabolic reactions inside root cells, etc. Within certain limits, high temperature facilitates all these criteria causing greater absorption of water.

6. Influence of the concentration of the soil solution on the rate of absorption

EXPERIMENT: 200 ml each of the following molar solutions of NaCl is prepared from 1 M NaCl (dissolve 58·5 gm NaCl in 1000 ml water): 0·3, 0·2, 0·1, 0·05 and 0·025. The solutions are poured into five separate bottles of 250 ml capacity. The sixth bottle is filled with 200 ml of distilled water. Six equal sized uniform aged vigorously growing sunflower seedlings are selected and weighed separately. The soil from the roots is carefully washed. Better results will be obtained with seedlings grown in water cultures. One seedling is supported in each bottle immersing its root system well into the solution and the portion of the stem which passes through the mouth of the bottle is plugged with cotton. The bottles are kept in moderately strong diffused light. After 4 hours, the seedlings are taken out, washed and reweighed. Increase in water content of each seedling is determined by taking the difference of final and initial weights and is expressed as gm of water absorbed per gm fresh weight. A graph may be suitably plotted.

OBSERVATION: Absorption of water decreases with increase in molar concentration of the solution.

INFERENCE: Water absorption by roots becomes difficult or impossible if the concentration of soil solution is higher than that of the protoplasm of the root cells. Concentration of soil solutions depends upon the presence of soluble substances in the soil. If water absorption by roots is due to osmotic forces, then the D.P.D. of water in the soil solutions should be lower than that of the root cells.

7. **Influence of substratum-air on the rate of absorption of water**

EXPERIMENT: Two suitable wide-mouthed bottles are filled with distilled water. Air-bubbling arrangements are made in one bottle for aeration of the substratum. For convenience, instead of bubbling, fresh *Hydrilla* plants may be introduced into the bottle for oxygenation of the substratum. Five healthy uniform-sized *Phaseolus* seedlings are weighed and put in each bottle so that the roots are completely immersed in water. Both the bottles are kept under bright sun light. After 4 hours seedlings are taken out and reweighed.

RESULTS: Difference between the final and the initial weights in both the sets gives the amount of water absorbed by five seedlings in 4 hours. From this data the amount of water absorbed per gm of seedling per hour is calculated. Two bar graphs are drawn for comparison of results.

DISCUSSION: From the results it is clear that the aeration of substratum increases the water absorption by roots. When water is absorbed actively, some extra energy which is produced by metabolic activity of living cells is required. By utilisation of this energy water is thought to be pumped or forced into the xylem vessels. Thus water absorption process is partly dependent upon the respiratory activity of the root cells and is favoured by the availability of oxygen in the substratum. If aeration of soil is not sufficient, it will produce an inhibitory effect on the metabolism of root cells leading to less absorption of water.

8. **Experiment to demonstrate the influence of soil texture on water absorption**

EXPERIMENT: Three types of soil, viz., sandy, loamy (garden soil) and clayey (collected from tank bed) are sun-dried for 48 hours. 500 gm of each type of soil is taken separately in three small earthenware pots having holes at the bottom. Several *Phaseolus* or *Vigna* seedlings (10-day old) of uniform size are selected. The average water content of the seedlings is determined by taking fresh and dry weights of a few seedlings. Now, 10 fresh seedlings are transplanted in each pot giving equal spacing. Each pot is watered with 50 ml of water and kept in an open space. Care is to be taken to check evaporation from the soil surface. After 2 or 3 days the seedlings are taken out and the average water content is determined.

RESULTS: The difference between the final and the initial water content of the seedlings in each case indicates the amount of water absorbed by the seedlings growing in each pot. Maximum water content is obtained in the seedlings growing in loamy soil and minimum in sandy soil.

DISCUSSION: The best soil for the growth of a plant is that which has normally sufficient water and air holding capacities. Loamy soil satisfies this condition, as it contains equal proportions of sand, silt and clay. This soil possesses its good porous nature due to presence of sand while clay and silt increase its water-holding capacity. The sandy soil has the least, clay the maximum and the loam the intermediate relative values of field capacity. Hence maximum water is absorbed by the plants grown in loamy soil.

9. **Effect of promoter and inhibitor on the rate of water absorption:**

EXPERIMENT: To perform this experiment the following promoters and inhibitors may be conveniently used: IAA (0·01%), glucose (5%), malic acid (5%) as promoters and chloramphenicol (0·01%), DNP (1%) and mercuric chloride (1%) as inhibitors.

Three wide-mouthed bottles are taken. One is filled with any promoter solution, the second with any inhibitor solution and the third with tap water which acts as a control. Five suitable seedlings of uniform size and age are selected for each set and fresh weight is determined. These are then put in each bottle. The bottles are left for 4 hours and the final weight of each set of seedlings is taken again.

RESULTS: The difference between the final and the initial weight in each case gives the amount of water absorbed by a set of five plants in 4 hours. The amount of water absorbed per gm fresh weight of the seedlings per hour is calculated and graphically represented.

DISCUSSION: From the results it is clear that promoters increase absorption of water over control while inhibitors decrease it. IAA increases water absorption by softening the cell wall, thus permitting a passive osmotic uptake of water. Glucose and malic acid, being respiratory substrates, increase active absorption of water by increasing the rate of respiration. DNP being an uncoupler of oxidative phosphorylation, hampers the active absorption process. Mercuric chloride being a toxic substance, inhibits respiration of root cells thereby inhibiting active absorption. Chloramphenicol is a metabolic inhibitor and retards general metabolism particularly protein and RNA synthesis. Studies with promoters and inhibitors show that living cells and their metabolic activities take some part in water absorption process.

10. **Experiment to demonstrate the translocation of water or ascent of sap.**

EXPERIMENT: In this experiment both ascent of liquid through the stem and the effect of transpiration on the rise of liquid can be shown. The leafy shoot of a suitable plant having two branches is taken and kept in water. From one of the side branches all leaves are removed and the stem is cut partly with a knife from the apical region. A narrow glass tube is fitted with this end with the help of a rubber tubing. The glass tube is filled with water and immersed in a beaker containing mercury. The tube is clamped properly (Figure 8). All connections are made air-tight. Now the water is continuous throughout the whole system extending from the mesophyll cells of the leaf, down through the xylem vessels of the stem upto the end of the glass tube immersed in the beaker. The whole set-up is kept for 2 to 3 hours under bright sun light.

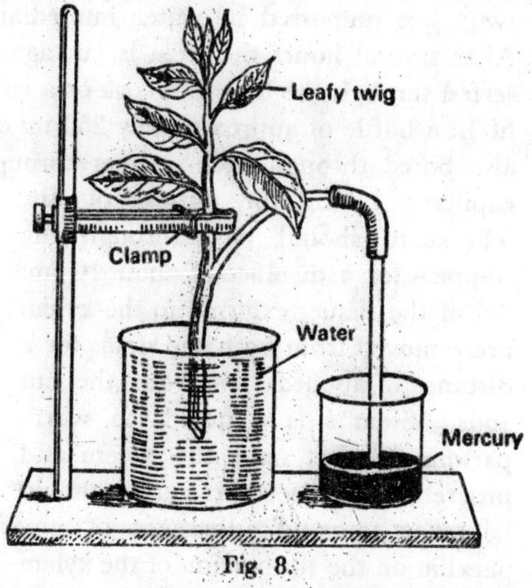

Fig. 8.

OBSERVATION: After the desired time mercury is found to rise in the narrow glass tube.

INFERENCE: Under these conditions as water molecules are lost by transpiration from the leaves, an upward pull develops causing mercury to move up into the glass tube. The mercury column remains intact due to cohesive forces.

N.B. The rate of ascent of water may also be determined from this experiment by taking data on rise of mercury in the glass tube per unit time.

11. **Experiment to demonstrate that xylem is the path of conduction of water:**

(*i*) EXPERIMENT: Leafy twigs of *Vinca*, *Ervatamia* or any other suitable plant are immersed in red coloured eosin solution. After a few hours cross sections are made from different heights of the stem and observed under the microscope.

OBSERVATION: It becomes evident that the red colour is primarily confined to the xylem vessels.

INFERENCE: This proves that the xylem strands are the main path of upward translocation of water in plants.

(*ii*) EXPERIMENT: Two leaf bearing woody stems, on which the aggregate leaf area appears to be equal, are selected. The basal ends of the twigs are immersed in water immediately after cutting from the plant. After several hours the stem is cut again under water. Each stem is inserted through the hole of a cork or a rubber stopper of proper diameter to fit in a bottle of approximately 250 ml capacity. A second smaller hole is also bored through each stopper through which is inserted a short bent capillary tubing for passage of air. The stems should pass through the stoppers for a distance of about 10 cm. All of the tissues external to the xylem are removed from both the stems for a distance of about 3 cm from the cut ends. Stem *b* is coated with warm paraffin so that only the xylem and pith remain exposed. Precaution should be taken to avoid adherence of any paraffin on the tip portion of the xylem core. Stem *a* is also coated with paraffin in such a way that the tissues external to the xylem (i.e., at the point from where tissues have been removed) remain exposed (Figure 9). Each bottle

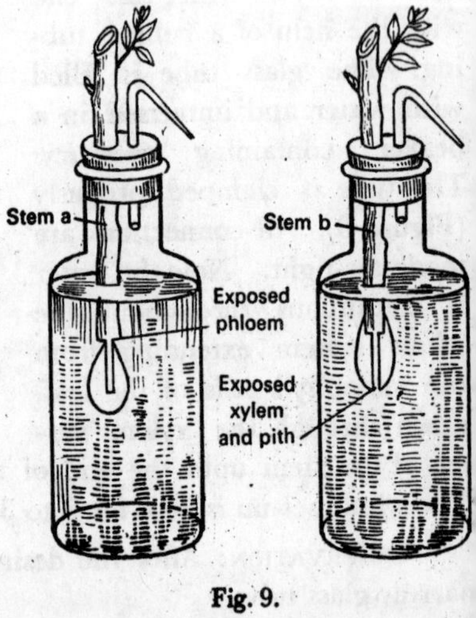

Fig. 9.

is nearly filled with a known volume of tap water and each is stoppered with one of the stoppers through which passes one of the stems. Both the sets are kept in bright sun light.

OBSERVATION: The behaviour of the leaves is observed from time to time until the leaves of one twig show sign of wilting and drying. It is observed that the stem *a* shows such sign after some time while the stem *b* does not. The volumes of water in both the bottles are measured. The difference between the initial and the final volumes gives the amount of water absorbed by the stems. Absorption of water is higher in case of stem *b* compared to stem *a*.

INFERENCE: From this experiment it can be inferred that water is absorbed and translocated mainly through the xylem and not through the phloem.

(*iii*) EXPERIMENT: A forked twig of *Gardenia* or *Vinca* or china rose is taken and its main axis is left intact. From the left hand branch all the cortical tissues are removed from a narrow ring. From the right hand branch the central woody strand of xylem is very carefully severed from a narrow zone leaving the cortical tissues undamaged. The lower end of the main axis is immersed in water coloured with eosin. After 3 or 4 hours observations are made.

OBSERVATION: It is observed that the leaves of the right hand branch have considerably wilted and no eosin could be detected above the ring of incision. The leaves of the left hand branch are more or less unaffected and eosin could be easily detected above the stripped zone.

INFERENCE: Severence of vascular strands stops upward translocation of liquids causing wilting of leaves. Removal of cortical tissues does not interfere with the conduction of water or liquid if the xylem remains intact.

12. Experiment to demonstrate that living tissues take little part in the upward translocation of water:

EXPERIMENT: Two suitable twigs of *Ervatamia* are taken and the end of one twig is dipped in boiling water for some time and the end of the other twig is dipped in a strong solution of $CuSO_4$ or $HgCl_2$ or any other toxic substance for the same period. After such drastic treatment which kills all the living cells of the treated regions, twigs are put in a beaker containing coloured eosin solution. After about 4 hours observations are made.

OBSERVATION: At the end of the experiment eosin can be detected in the xylem strands through out the stem and leaf of both the twigs when cross sections of stem or leaf are observed under the microscope. There is very little wilting of leaves in each case.

INFERENCE: This experiment indicates that living cells take very little part, if any, in upward translocation of liquid in vascular plants.

CHAPTER VI

TRANSPIRATION

Transpiration is a term applied generally to the loss of water in the form of vapour from leaf tissue. The process is in principle one of evaporation and diffusion, but not evaporation from perfectly open and free surface. The process of transpiration is controlled by the activity of living plant tissue. The evaporation is a physical process which can occur from both living and nonliving surfaces.

On absorption from the soil by roots, water is translocated via the xylem tissue to the mesophyll cells of the leaves. The loose arrangement of living thin-walled cells, having large intercellular spaces, provides congenial conditions for the evaporation of water from cell surfaces through innumerable microscopic pores called *stomata.* Stomata provide an uninterrupted path from the interior of the leaf to the external environment. In addition to *stomatal transpiration* mentioned above, water is lost also as a vapour directly from the surfaces of leaves and herbaceous stems, and through *lenticels* (small openings in the corky tissue covering stems and twigs). The former is called *cuticular transpiration* because it involves diffusion of water vapour directly through the *cuticle* and the latter the *lenticular transpiration*. The loss of water in the form of liquid from the plant tissue is called *guttation* or *secretion.*

The stomatal transpiration is controlled by opening and closing of stomata. The movement of stomata which causes the opening or closing of stomatal pore, is due to the changes in turgidity of the guard cells. When the guard cells are turgid the stoma opens but when they are flaccid stoma closes. Thus the size of the stomatal aperture is related to the turgidity of the guard cells. The walls of guard cells towards the pore are more thick and inelastic. When the turgidity increases in the guard cells, the pore widens and becomes open. The reverse event occurs when the guard cells lose turgidity. The turgidity of the cells is dependent upon its osmotic pressure capacity. The changes in the osmotic pressure of the guard cells are again dependent upon the position of equilibrium of the reversible chemical conversion of starch to sugar which is controlled by the enzyme *phosphorylase.* This equilibrium is determined by the effects of light, pH in the guard cells, temperature, water content and CO_2 concentration in guard cells, etc. Starch is converted into sugar when in light, CO_2 is used up in photosynthesis and pH remains high (7·0). Osmotic pressure increases in the guard cells and endosmosis of water into guard cells takes place. The guard cells become turgid and stoma opens. In dark sugar is converted into starch when CO_2 accumulates and pH decreases (5·0). Decrease of osmotic pressure in guard cells takes place and exosmosis of water from them results in flaccidity of guard cells closing stomata.

Recently it has been shown that potassium ions are actively absorbed in expense of energy by the guard cells and osmotic pressure increases resulting in stomatal opening in light. Necessary ATP molecules are supplied during photosynthesis by guard cell chloroplastids. In dark these potassium ions are pumped out resulting in decreased osmotic pressure and subsequent closure of stomata. Moreover, in darkness, air containing CO_2 apparently inhibits the accumulation of potassium ions in guard cells, thus the stomata do not open. In light, on the other hand, CO_2 conc. is decreased, potassium ions are accumulated at an enhanced rate and stomata open.

Both the internal factors of plant itself, like stomatal frequency and behaviour, plant structures, and the external factors of the environment like light, temperature, relative humidity of air, wind velocity, available soil water, atmospheric pressure, and the nature of gas into which the water vapour diffuses, regulate transpiration.

The effects of transpiration on the plants are based upon the following points: (1) cooling effects produced due to the evaporation of water from plant surface, (2) exit of water vapour through stomata or epidermis by diffusion, (3) absorption of water by roots, translocation of water by xylem and supply of water to the leaves and other parts, (4) salt absorption and translocation, (5) reduction of water content in plant and soil, and (6) increase in the humidity of air near the plants.

In the following experiments, the various aspects of transpiration and its regulation have been taken into consideration.

1. Determination of stomatal frequency (or the number of stomata per unit area) of a leaf

EXPERIMENT: In order to determine the dimension of stomata and the area of field of vision, the ocular is first standardized with the help of a stage micrometer. One stage micrometer division is generally 10μ (it is indicated on the stage itself). The length of the scale of the stage micrometer is equal to 1 mm and 1 mm is equal to 1000μ. For standardization of the ocular micrometer, it is first placed inside the eye-piece and the stage micrometer on the stage of a microscope at a particular magnification. Now it is observed how many divisions of the ocular coincide exactly with those of stage micrometer. Three such readings are taken and the average is determined. The value of one ocular division is calculated as follows:

If X stage divisions are equal to Y ocular divisions, then one ocular division becomes equal to $10X/Y\mu$ (1 stage division $=10\mu$). Standardization in both low and high power of microscope is separately done.

The diameter of the field of vision of the microscope is determined with the help of the ocular or the stage. The area of the field of vision is calculated from the formula πr^2, where r is the radius of the field of vision, i.e., half of the diameter (convert μ into centimeter and calculate area of the field of vision in square centimeter).

Epidermal peelings from the lower surface of *Basella* leaf are mounted on a slide and the number of stomata within the field of vision is counted. Three such readings are taken by moving the slide. Such readings are taken from the apical, middle and the basal portions of the leaf and the average number is calculated.

RESULTS: The number of stomata within the area of field of vision is determined and from it the number of stomata per square centimeter is calculated. This gives the stomatal frequency of the leaf.

DISCUSSION: Stomatal frequency is defined as the number of stomata per unit area of the leaf surface. Different species of plant have different stomatal frequencies and this varies with environment and age of the leaves. In leaves stomata may occur in both upper and lower surfaces. In woody plants with dorsiventral leaves they are located mainly on the lower epidermis and in herbaceous plants with isobilateral leaves they occur on both the surfaces, though more abundant on the lower side. In an individual leaf stomata are more numerous near the apex and minimum near the base. the middle region having an average distribution. Stomatal frequency generally ranges between few thousands to a hundred thousands. In determining stomatal frequency of an isobilateral leaf the stomata of both the surfaces should be taken into consideration.

2. Measurement of stomatal pore

There are several methods for determination of the state of stomatal opening. A few important methods are given below.

(*a*) LLOYD'S METHOD: By this method stomatal pore of leaves can be measured by using ocular (standardized) and the microscope. The epidermal peelings of leaf of a suitable species are taken and immediately fixed by immersing in hot alcohol. The alcohol fixes the stomata preventing any further movement of the guard cells. The peelings are examined under the microscope and the opening of the guard cells is then measured by a standardized ocular. The area of the pore is determined. Since the area of the pore is somewhat ellipsoidal in nature, the area could be very approximately taken as equivalent to $(\pi/4)\ (a \times b)$, where a and b represent the length and the breadth of the pore, and $\pi = 3{\cdot}142$.

(*b*) IMPRESSION METHOD: The leaf is taken from a healthy potted plant and its lower surface is smeared with **Durofix or Collodian** (1 gm of pyroxylin in 6 ml of alcohol and 20 ml of ether). It is kept for some time. After about half an hour the Durofix or Collodian is dried up into a thin papery film. The film is now stripped off and the impressions of the stomata in such films can be observed under a microscope. The exact area of the stomatal opening is calculated by the above method.

(*c*) DARWIN'S POROMETER METHOD: Darwin's porometer is a useful apparatus for following the changes in the stomatal aperture, i.e., the degree of opening of stomata. This is an indirect method for knowing the degree of opening of stomatal aperture. The principle of the method is that the water vapour, which comes out of the stomatal pores, produces a pressure on the level of water column in a vertical tube pushing the level down at a rate at which water vapour comes out (Figure 10). The figure shows a T-tube with two side arms A and B

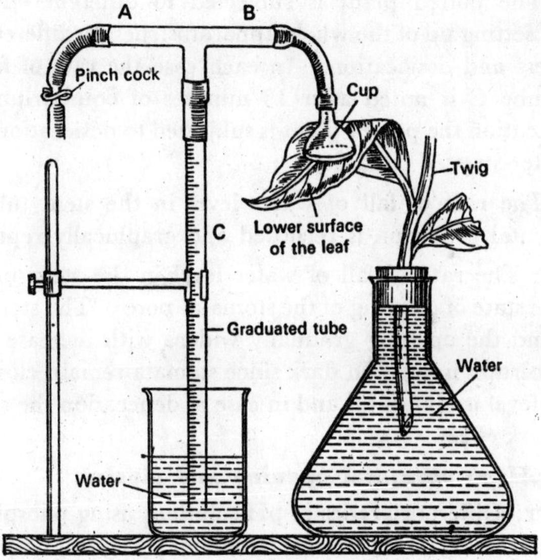

Fig. 10.

at top and a graduated vertical tube C. The arm A is connected to a rubber tube on which a pinch cock is fitted to cut off the flow of air when desired. The arm B is connected to a cup by a rubber tubing. The mouth of the cup is attached to the lower surface of a leaf by some adhesive or vaseline. Care should be taken so that adhesive or vaseline does not adhere to the leaf surface thereby closing the stomata. The fittings must be perfectly air-tight. The lower open end of the stem tube C is immersed in a beaker containing water.

At the start, the pinch cock is opened and the air is sucked out by mouth so that water level rises in the stem tube C. The level of water is adjusted at the desired height and the pinch cock is closed. If the stomata on the leaf surface within the cup are open, then air will enter through the leaf surface via leaf tissue and will come out of stomatal pore along with transpired water vapour. This air will move via the cup chamber, rubber tube and finally it will exert pressure on the level of water in the stem tube pushing it down at a rate at which air is coming out of the stomatal pore, depending upon their degree of opening. The experiment can be done under different environmental conditions to observe their effects on the degree of opening of stomata.

3. Determination of changes of stomatal opening in light, dark and under desiccation

EXPERIMENT: Darwin's porometer is used in this experiment (Refer Expt. 2. c). The potted plant is subjected to different environmental conditions after setting up of the whole apparatus, such as different intensities of light, darkness and desiccation. In each case the rate of fall of water level in stem tube C is noted after 15 minutes of equilibrium. For the purpose of desiccation the potted plant is subjected to desiccation by keeping the plant in water-stressed condition.

RESULTS: The rate of fall of water level in the stem tube C under each environmental condition is recorded and graphically represented.

DISCUSSION: The rate of fall of water level in the stem tube C is an indication of the state of opening of the stomatal pore. The stomata remain open in light and the opening gradually widens with increase in light intensity up to a certain limit. In dark since stomata remain closed, the rate of fall of water level is almost nil and in case of desiccation the rate of fall is similar to the dark effect.

4. Effect of pH on stomatal opening and closing

EXPERIMENT: Different grades of pH solutions using phosphate buffers are prepared by mixing 0·1M disodium monohydrogen phosphate (Mol. wt. 142·16) and 0·1M monosodium dihydrogen phosphate (Mol. wt. 156) solutions in the following proportions:

Na_2HPO_4	:	$NaH_2PO_4, 2H_2O$	pH
10	:	0	9·00
8	:	2	7·20
6	:	4	6·75
4	:	6	6·50
2	:	8	6·00
0	:	10	4·00

If Na_2HPO_4 contains 2 mols of water of crystallization, then the molecular weight will be 178·16. Sodium phosphate buffer may be substituted by potassium phosphate buffer (Mol. wt. of K_2HPO_4=174·20 and KH_2PO_4=136·20. If water of crystallization is present, it should be taken into account.)

Six watch glasses are taken and 2 ml of each buffer solution is poured in each watch glass separately. Another watch glass containing 2 ml of distilled water serves as control. Epidermal peelings from suitable leaf are taken and put in each watch glass. After 10 to 15 minutes the peelings are taken out and observed under the microscope. The area of stomatal aperture is measured as in Expt. 2a. At least three such readings are taken for each set and average area of the pore is determined.

RESULTS: Results are plotted graphically taking pH as abscissa and area of the pore as ordinate. The pH at which the area is maximum is noted from the graph.

DISCUSSION: The extent of stomatal opening is greatly influenced by pH. Stomata remain closed in acidic pH whereas in alkaline pH stomata open. At higher pH (pH 7 to 7·20) and in presence of enzyme and inorganic phosphate, starch is converted to glucose-1-phosphate. At lower pH (pH 5) the synthesis of starch takes place from glucose-1-phosphate. Since starch is osmotically inactive and glucose osmotically active, the above reaction shows the possible explanation of the effect of pH on stomatal opening and closing.

N.B. The effect of temperature on stomatal opening and closing may also be studied. Under suitable environmental conditions, with the rise of temperature up to 30°C. opening of stomata increases.

The effect of potassium ions on the opening and closing of stomata may be stuided by taking different concentrations of KNO_3 and putting the epidermal peelings in these solutions under bright sunlight.

The effect of some metabolic inhibitors like DNP, cyanide, etc. may also be examined to show that the mechanism of opening and closing of stomata is controlled by the metabolic activity of the guard cells.

5. Determination of the area of leaves by different methods

(*a*) BY POLAR PLANIMETER METHOD: The Polar Planimeter was invented by Professor Amesler in 1856. This is a simple instrument for determination of areas of unspecified surfaces. It is a very useful apparatus as areas could be read out from readings in a small vernier attached to the instrument (Figure 11). It consists of two principal parts, a tracer arm having tracing point and a carriage with a measuring wheel, and the pole arm attached to the pole, around which the instrument revolves.

The main operation of the apparatus in determining the area of a leaf is to place the leaf on a paper or on a board (outline of the leaf may also be drawn on it), to keep the point of the instrument near one margin of the

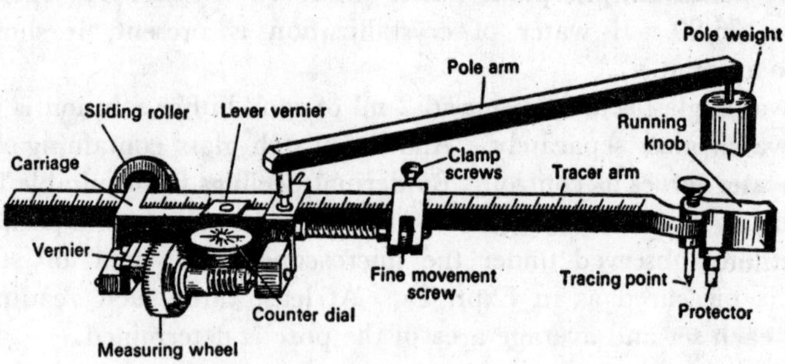

Fig. 11.

leaf and to pass it around the margin of the outline. The result is those computed from the reading of the graduated measuring wheel and the counter dial which registers the number of its revolutions. The measuring wheel is with a vernier, vernier unit is 10 sq. mm.

DIRECTION FOR USE:

1. Determination of vernier unit:

   First coincide the zero of the measuring wheel with the zero of the vernier scale. Find out the number of divisions of the vernier scale that coincides with that of the measuring wheel.

   Now, 10 vernier divisions = 9 divisions of the measuring wheel.

   ∴ 1 „ „ =9/10 or 0·9 „ „ „

   ∴ Vernier unit =1 − 0·9=0·1 sq. cm.

   1 small division of the measuring wheel =1 sq. cm. and

   1 division of the counter dial =100 sq. cm.

   ∴ Total reading =(Counter dial reading ×100) +

   + (Measuring wheel reading ×10) +

   + (Vernier reading ×0·10) sq. cm.

   (If the reading in the measuring wheel is below 1, say 8 divisions, reading is to be taken in decimal, i.e., 0·8).

2. Before taking actual measurement of area from the outline drawn on a paper, zero of the lever vernier should be fixed at 35 mark of the tracer arm.

(*b*) BY WEIGHING METHOD: A leaf is placed on a card board or on a century board and its outline is drawn. The leaf area on the board is accurately cut and weighed. Now one square centimeter area is cut out from the board and is weighed. If weight of the area drawn on the board is X gm and that of one square centimeter is Y gm, then the area of the leaf is X/Y sq. cm.

(*c*) BY GRAPH PAPER METHOD: This is the most convenient method which is generally used in laboratory experiments. The leaf is placed in a millimeter graph paper and its outline is drawn. The total number of large square blocks (one large square block is equal to one square centimeter) is counted. The number of small squares (one small square is equal to one square millimeter) in the remaining area of the outline of the leaf is then counted and converted into square centimeter. All the values are added in order to get the total area of the leaf.

6. Determination of the percentage of total stomatal aperture in relation to the area of the whole leaf

EXPERIMENT: Stomatal frequency (X) of the experimental leaf is determined as in Expt. 1 and the area of the leaf (Y) is found out as in Expt. 5c. Now the average area (Z) of the stomatal pore is determined as in Expt. 2.a.

RESULTS: The product of X (stomatal frequency) and Y (area of the leaf) gives the total number of stomata. This product when multiplied by Z (area of the aperture) gives the total area of the stomatal aperture of the leaf. The percentage of the area of the stomatal aperture in relation to the entire leaf area is calculated.

DISCUSSION: Stomatal frequency varies with different species of plants, age of leaf of the same species and the conditions under which it is grown. When there are a few stomata per unit area of the leaf, they are usually large, if there are many they are generally small. Calculations of stomatal areas as percentages of leaf areas, however, show no constant relation. The range is not very great and is usually between 0·4 to 0·2 percent of the leaf area calculated on the basis of one surface in case of dorsiventral or both the surfaces in case of isobilateral leaves.

7. Demonstration of law of diffusion through small pores

EXPERIMENT: The law of diffusion through small pores can be studied by taking equal amounts of alcohol in two wide-mouthed bottles. The mouths of the bottles are covered with two polythene sheets having holes of different diameters. Care should be taken that the holes are more or less widely spaced. The rate of evaporation is measured by weighing the bottles from time to time. The total diameter and perimeter in centimeter and area in square centimeter of all the pores are separately determined in the two sheets.

RESULTS: The area of the pores and the perimeter of the pores in each case are determined by applying the formulae $\pi d^2/4$ and $\pi \times d$ respectively, where d is the diameter of the pore. The mean change in weight per unit time per unit area and per unit linear dimension of the pore (perimeter) gives the rate of evaporation through pores of different diameters. The results are entered in tabular form.

DISCUSSION: In quiet air the rate of diffusion is more nearly proportional with the linear dimension, i.e., the perimeter of the pores than to their areas. For this reason the rate of diffusion per unit surface of pore is higher in the smaller pores. If the pores are uniformly scattered over a surface, the actual open area is highly reduced but there is slight reduction in diffusion rate. Thus the diffusion through a multiperforate septum is higher as compared with that of a single opening with an area equal to the aggregate area of the pores. Closely spaced pores bring interference in diffusion due to overlapping of humidity layers. It has been estimated that if pores are over 8 or 10 times diameter apart, interference is minimum and each pore allows for its maximum diffusion. Interestingly, stomata are usually further apart than 8 times of their diameter. This is why the rate of diffusion through stomatal apertures is high even though the total area of stomatal opening is small. Although, the open area of the stomatal pore only represents 1–2% of the total leaf area, the diffusion of water vapour through the pores often exceeds 50% of that evaporating from free water surface.

N.B. In order to establish the degree of correlation between (*i*) the rate of evaporation and the perimeter of pores and (*ii*) the rate of evaporation and the area of pores, results are tabulated and subjected to statistical analysis for degree of correlation (see Appendix I).

8. Demonstration of the phenomenon of transpiration

(*a*) BELL JAR METHOD (qualitative):

EXPERIMENT: A small healthy potted plant is taken and its soil is covered with a polythene sheet to check evaporation from the soil surface. The stem and leaves remain uncovered. The plant is now placed under a bell jar and the rim of the bell jar is sealed with vaseline. The set-up is kept in a lighted place. Another set-up is similarly maintained where all the leaves from the plant are removed previously.

OBSERVATION: In the first set-up droplets of water are seen on the inner walls of the bell jar. In the second set-up there is no condensation of water vapour on the inner wall of bell jar.

INFERENCE: When leaves transpire water, the water vapour comes out and after saturating the inner atmosphere of the bell jar it condenses in the form of droplets on the inner walls. In case where all the leaves are removed from the plant before covering with bell jar, no condensation of water vapour takes place because the transpiring organs, that is leaves, are absent.

N.B. Instead of removing the leaves, vaseline may be applied on both the surfaces of the leaves of the plant to prevent transpiration.

Effects of the presence of some poisonous gases (CO, CO_2, HCl vapour, etc.) on transpiration may be studied by introducing the gas into the bell jar before covering the plant. The effect of light and darkness may also be studied by this method.

(*b*) COBALT CHLORIDE PAPER METHOD (qualitative):

EXPERIMENT: A few pieces of filter paper are soaked in 5% aqueous cobalt chloride ($COCl_2$) solution for a few minutes. The pieces are then taken out, excess solution is removed by hanging and pieces are dried in hot air oven at 60°C. These are then cut into squares (2×2 cm) and kept in a desiccator containing fused $CaCl_2$. The original colour of the wet papers is pale pink, when dried to a standard uniform shade, the colour becomes intense blue.

Now a piece of cobalt chloride paper thus prepared is placed at the lower surface of a leaf of a potted plant and the paper is covered with glass slides from both surfaces of the leaf. The slides are made air-tight with vaseline.

OBSERVATION: After some time the blue colour of the cobalt chloride paper turns pink.

INFERENCE: As water vapour comes out of the stomatal pores (if they are open) it moistens the cobalt chloride paper changing its colour to pink. The treated blue dry paper is thus a moisture detector, turning pink when left in the air or when placed in contact or near an evaporating surface.

(*c*) BY CONICAL FLASK-WATER-OIL-LEAF METHOD (quantitative):

EXPERIMENT: A 250 ml conical flask is filled up to its neck with water and a suitable petiolate leaf (cut under water in order to maintain continuity of water column) is inserted within the conical flask in such a way that its petiole remains under water. A sinker may be used to keep the leaf in position. Now oil is poured on the surface of water to check evaporation from water surface. The whole set-up is weighed and allowed to stand in open air or in a well ventilated room having sufficient light. After 2 hours the set-up is reweighed.

RESULTS: The difference between the first and the second weights gives the amount of water transpired by the leaf in 2 hours. The results may be expressed in per unit time.

DISCUSSION: The loss in weight is due to loss of water in the form of water vapours through the leaf. Since the free evaporating surface is checked by oil film, the sole loss of water is only due to transpiration through the leaf.

(*d*) HANGING LEAF METHOD:

EXPERIMENT: Two healthy leaves of a suitable plant are selected. One of the leaves is smeared with vaseline on both the surfaces and to the petiole to check transpiration. The second leaf is left as such. The leaves are then weighed and kept hanging in bright sunlight for an hour. These are then taken out and reweighed.

RESULTS: After one hour it is seen that the first leaf remains turgid and fresh while the second leaf wilts. In the first leaf the difference between the initial and the final weights are negligible, but in the second leaf considerable loss in weight takes place.

DISCUSSION: Results indicate that the first leaf in which transpiration is checked, the loss of water is insignificant and the leaf remains turgid and fresh. While in the second leaf loss of water due to transpiration takes place in appreciable quantity and the leaf wilts.

9. Determination of transpiration index

EXPERIMENT: Transpiration index is defined as the ratio of the time (in seconds) required for a standard change of colour of dry cobalt chloride paper over a free evaporating surface (S) and the transpiring surface (E) and can be expressed as follows:

$$\text{Transpiration index} = \frac{S \times 100}{E}.$$

(The index is multiplied by 100 to obtain the loss of water from the leaf surface as a percentage of evaporation from free water surface).

Two equal pieces (2×2 cm) of cobalt chloride papers are taken. One is attached to the lower surface of a dorsiventral leaf as in Expt. 8.b and another is placed on a wire net which is kept over a petridish containing water. Time taken in seconds for a standard colour change of the cobalt chloride papers in both the cases is noted.

RESULTS: Transpiration index is calculated from the above mentioned formula.

DISCUSSION: Transpiration index gives an indication of the relative efficiency of the rate of transpiration with that of physical evaporation. It varies from species to species, with age of leaves and with different environmental conditions.

N.B. **Transpiration index of upper and lower surfaces of a leaf, in** light and dark, in young and old leaves and in different ecological types of leaves may be compared.

The participation of conducting tissues in transpiration may also be studied from this experiment by separately removing xylem and phloem tissues.

10. **Determination of the rate of transpiration**

(*a*) COBALT CHLORIDE PAPER METHOD (qualitative):

Two pieces of cobalt chloride papers are affixed on both the surfaces of a dorsiventral leaf as in Expt. 8.b. The length of time required for the papers to turn pink indicates the rate of transpiration from the two surfaces of the leaf.

(*b*) CONICAL FLASK-WATER-OIL-LEAF METHOD (quantitative):

EXPERIMENT: Two dorsiventral leaves of equal and comparable sizes are taken. The upper surface of one leaf and the lower surface of the other are smeared perfectly with vaseline. The experiment is then set-up as in Expt. 8.c.

RESULTS: The amount of water transpired per unit time divided by the area of the leaf (determined by graphical method as in Expt. 5.c) gives the rate of transpiration per unit time per unit area of the leaf.

DISCUSSION: The rate of transpiration is higher in lower surface than in upper surface because stomata are more abundant in lower surface than upper surface which is cuticularised. The rate of transpiration from the lower surface indicates stomatal transpiration and that from the upper surface indicates cuticular transpiration.

N.B. The rate of transpiration may be correlated with (*i*) stomatal frequency, (*ii*) a single stoma (amount of water transpired per unit time divided by the total number of stomata of the leaf), (*iii*) area of stomatal aperture (amount of water transpired per unit time by the total stomatal aperture), (*iv*) number of xylem vessels present in the petiole (a cross section of the petiole may be cut for estimation of total number of xylem; the total area of xylem vessels as determined from the formula πr^2 and correlated with the rate of transpiration), (*v*) simple physical evaporation (rate of evaporation of water per unit time per unit area of a petridish is to be calculated taking area of petridish to be πr^2) and this may be compared with the rate of transpiration per unit area of the stomatal aperture.

(*c*) POTOMETER METHOD (quantitative):

The rate of transpiration (expressed in gm per hour per square centimeter of the leaf surface) can be measured with the help of an apparatus known as *potometer*. Determination of the rate of transpiration by different potometers, excepting Garreau's, is an indirect one where both absorption and transpiration have been taken into consideration. Potometers are designed on the principle that the rate of transpiration is nearly proportional to the rate of absorption of water by the plant, although it is not the general rule. There are many types of potometers, all working on the same principle excepting Garreau's which gives a direct measurement of transpiration. The potometers described below are generally used in laboratory experiments.

(*i*) DARWIN'S POTOMETER:

DESCRIPTION: This apparatus consists of a short glass tube from which a side tube bends upward ending in an open mouth (manometer tube) into which a plant twig is inserted through a cork. The upper open mouth of the main tube is closed by a cork. The lower end of the tube is also fitted with a cork through which passes a long graduated (in ml) capillary tube. The end of the capillary tube dips in a beaker containing water (Figure 12).

Fig. 12.

EXPERIMENT: At the beginning of the experiment, water is filled up in all the tubes maintaining continuity. A twig (cut under water) having some leaves is inserted through the cork of the side tube. All joints should be air-tight. A small bubble is introduced through the lower end of the capillary tube. As the transpiration occurs from leaves of the twig, water is absorbed by the twig from the side tube and this produces a suction force which sucks up water from the capillary tube. As a result, air bubble within the capillary tube gradually moves upward.

RESULTS: The rate of upward movement of air bubble is recorded by noting the height by which it ascends in a given time. The rate of transpiration is expressed as the ml of water transpired per minute per unit area of the leaves.

DISCUSSION: See Expt. 10c (*iii*).

(*ii*) GANONG'S POTOMETER:

DESCRIPTION: This apparatus is most suitable for determination of the rate of transpiration. It consists of a narrow graduated horizontal limb which holds two vertical wide-mouthed tubes—one of which is fitted with a cork through which passes a leafy twig (cut under water) while the other acts as a reservoir which is fitted with a stopcock in the connecting tube to control water supply. The other end of the horizontal limb bends at right angle and at the opposite side of the vertical wide-mouthed tube (Figure 13).

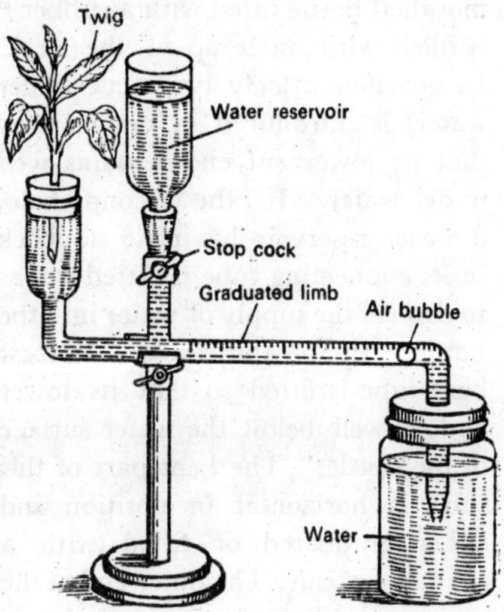

Fig. 13.

EXPERIMENT: At the outset, water is filled in all parts of the tube by opening the stopcock of the reservoir. A leafy twig is inserted through the cork of the vertical tube. Keeping the stopcock closed, all the connections are made air-tight. The twig is then allowed to transpire for a short period as a result of which water moves upward in the lower bent tube which dips in a beaker of water. An air bubble is now introduced through the lower end of the bent tube and its initial position is noted. The apparatus is fixed in this position with a stand and clamp. As water moves due to transpiration, the bubble also moves in the horizontal graduated limb. When the bubble comes to one end of the limb, it can be pushed back to its original position by opening the stopcock of the reservoir and the experiment can be repeated several times.

RESULTS: The rate of movement of the bubble in the horizontal tube is followed by measuring the distance it moves along the graduated scale within a certain period of time. This is considered as proportional to transpiration rate.

The rate of transpiration can be obtained by the following calculations:

Initial position of the bubble on the scale— X cm.
Final position of the bubble after a given time (t)— Y cm.

Therefore, the distance traversed by the bubble in time t is equal to (Y−X) cm. Now, volume of water transpired in time t is equal to πr^2 (Y−X) ml, where r is the internal radius of the bore. Thus the volume of water transpired by per unit area of the leaves of the twig per unit time is equal to $[\pi r^2 \{(Y-X)/t\} \times \text{area of leaves}]$ ml.

DISCUSSION: See Expt. 10c (*iii*).

(*iii*) FARMER'S POTOMETER:

DESCRIPTION: It has the same principle as that of Ganong's potometer and is only a modification of the latter. The apparatus consists of a wide-mouthed bottle fitted with a rubber stopper having three holes. The bottle is filled with water up to the neck. In one hole a leafy twig (cut under water) is introduced in such a way that its lower cut end remains well under water. In the second hole, a water reservoir having a stopcock in its connecting tube is fitted so as to control the supply of water into the bottle. In the third hole, a narrow bent tube is fitted so that its lower end is well below the water surface of the bottle. The bent part of this tube is horizontal in position and either graduated or fitted with a centimeter scale. The outer end of the horizontal tube is again bent downward and immersed in water (Figure 14).

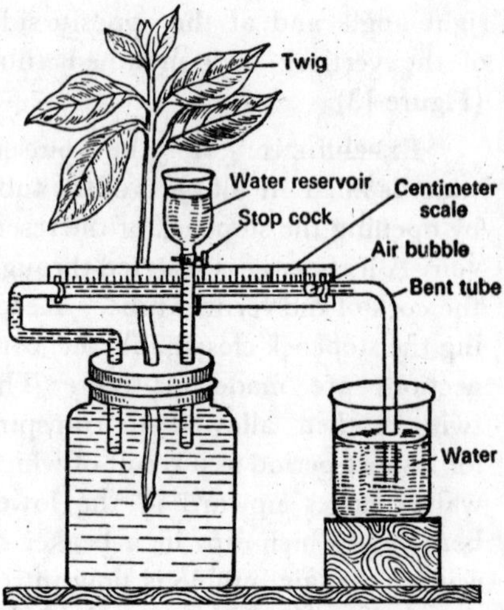

Fig. 14.

EXPERIMENT: At the beginning, the bottle is filled with water by opening the stopcock of the reservoir up to the mouth and also in the horizontal tube. The stopcock is then closed. An air bubble is now introduced into the outer bent end of the horizontal tube in the usual manner. As transpiration occurs from the leaves of twigs, water is absorbed from the lower cut end of the twig and a suction force is produced which sucks water from the horizontal tube. As a result air bubble moves inward. The rate of movement of air bubble is noted, which is considered proportional to the transpiration rate.

RESULTS: As in Expt. 10c (*ii*).

DISCUSSION: With the help of potometers (Darwin's, Ganong's & Farmer's) the rate of transpiration can be easily determined. This is an indirect method because the idea of this experiment is based on the assumption that the amount of water transpired by a twig is almost equal to the amount of water absorbed. In fact this does not generally happen in plants and the amount of water absorbed by a plant is not fully transpired but some amount is retained by it for its normal metabolic functions. Hence, the potometer methods described above for determination of rate of transpiration are not fully accurate.

(*iv*) GARREAU'S POTOMETER:

DESCRIPTION AND EXPERIMENT: This potometer is conveniently used for quantitative measurement of differential rates of transpiration from both upper and lower surfaces of a leaf. It consists of two wide-mouthed cups which are placed face to face keeping a widely expanded dorsiventral leaf of a twig from a potted plant in between them. Before keeping the leaf in between the cups, some anhydrous $CaCl_2$ contained in two small vials are weighed and placed in both the cups. The ends of the cups are closed with corks through which two mercury manometers are connected in order to keep the vapour pressure within the cups constant. The broad rims of the cups in contact with leaf surface is made air-tight by applying vaseline carefully. A change in vapour pressure within the cups shown by the manometers is indicative of the fact that either the connections are not air-tight or that all the vapour given out by the leaf surface is not being absorbed by the $CaCl_2$ within the vials. The whole arrangement is clamped vertically on a stand (Figure 15).

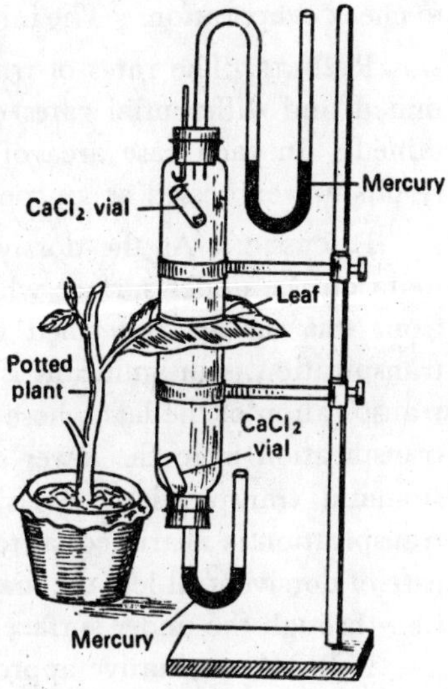

Fig. 15.

RESULTS: After a few hours, $CaCl_2$ vials are taken out and weighed again. The difference between the two weighings is a measure of the amount of water loss from the two leaf surfaces. As the areas are the same, this is also a quantitative measure of the differential rates of transpiration.

DISCUSSION: By Garreau's potometer method, not only the rate of transpiration but a comparative assessment of the rate of transpiration of upper and lower surfaces of a leaf, i.e., cuticular and stomatal transpiration respectively, can be simultaneously determined. This is a very convenient and accurate method for direct and quantitative measurement of the rate of transpiration in contrast to other potometer methods.

N.B. It is easily possible to set up potometer experiments under different environmental conditions such as at different temperatures or light intensities or relative humidities or wind velocities and to observe their effects on transpiration rate.

11. Quantitative determination of the differential rates of transpiration from the two surfaces of a leaf (cuticular and stomatal)

EXPERIMENT: This experiment is performed with the help of a conical flask-water-oil-leaf method as in Expt. 10b. Three dorsiventral leaves (betel or *Hibiscus* leaf) of almost uniform size and age are selected. Three conical flasks filled with water up to the neck are taken: (*i*) with untreated leaf, (*ii*) with leaf whose upper surface is thinly smeared with vaseline, and (*iii*) with leaf whose lower surface is similarly smeared with vaseline. Oil is poured in each flask to make a thin film on the water surface to check evaporation. The initial weights are noted separately in each case.

RESULTS: The rates of transpiration of all the three leaves are determined and differential rates of transpiration from either surface are obtained. In each case area of the evaporating surface is determined and results are expressed as gm per unit area per unit time.

DISCUSSION: As the dorsiventral leaves have a very few number of stomata on the upper surface which is cuticularized, the rate of transpiration from this surface is termed cuticular transpiration. That the cuticular transpiration is insignificant is clear when it is compared with the rate of transpiration of the leaf whose lower surface has been smeared. The rate of transpiration from the lower surface, having abundant stomata, gives the stomatal transpiration. It is generally found that the rate of stomatal transpiration is nearly equal to that of control leaf. This indicates that in case of dorsiventral leaf the major loss of water takes place through stomata, i.e., through the under surface of the leaf.

N.B. A qualitative approach regarding this experiment may be made by cobalt chloride paper method (Expt. 10a).

If a quantitative estimation of amount of water lost per stoma in a leaf is desired, the total number of stomata in both upper and lower surfaces is determined as also the total amount of water lost. The amount of water lost from the transpiring surface divided by the total number of stomata gives an approximate measure of the amount of water given out per stoma.

12. Simultaneous determination of the amount of water absorbed, retained and transpired by a plant

(*a*) BY DIRECT MEASUREMENT METHOD:

EXPERIMENT: This is a simple apparatus consisting of a wide-mouthed bottle with a graduated side tube (in ml) attached to its base through a cork. The mouth of the bottle is fitted with a cork through which is introduced a small leafy plant. The bottle is filled up with water and all connections are made air-tight (Figure 16). The evaporation from the side tube is prevented by pouring a thin film of oil on the exposed water surface. The level of water in the side tube is recorded. The whole arrangement is weighed and kept in bright sunlight. After 2 hours or so the level of water in the side tube is again recorded and the whole arrangement is reweighed.

Results: The difference between the initial and final levels of water in the side tube is a measure of the amount of water absorbed in ml by the plant and the difference between the first and the second weight evidently shows the amount of water transpired in gm from the leaf surface. The volume of water absorbed may be converted to gm by multiplying the density of water at that temperature from a standard temperature-density table (see Appendix II). Now the difference between the amount of water absorbed and the amount of water transpired indicates the amount of water retained.

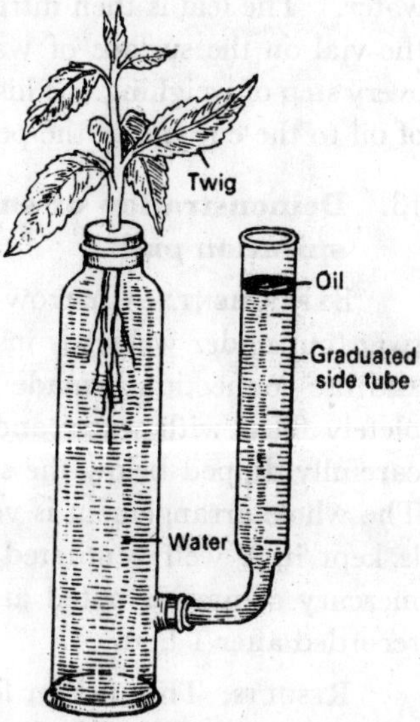

Fig. 16.

Discussion: See Expt. 12b.

(*b*) **By conical flask-water-oil-leaf method**:

Experiment: The above relation can also be experimentally determined by conical flask-water-oil-leaf method (Expt. 10b). The weight of the conical flask plus water plus oil is taken (W_1). Care should be taken to avoid excess addition of oil. There should be a thin film of oil covering the free evaporating surface. Now the petiole of a leaf from a suitable plant is carefully introduced through the open water surface by tilting the conical flask so that no oil adheres to the petiolar end (see N.B.). The weight of the conical flask plus water plus oil plus leaf (W_2) is noted. The whole arrangement is exposed to a well ventilated lighted place for 2-hours. The set-up is then reweighed (W_3). The leaf is taken out carefully so that no water or oil adheres to it. The weight of the conical flask plus water plus oil is again noted (W_4).

Results: Now, ($W_1 - W_4$) is equal to the amount of water absorbed by the leaf (*i*). ($W_2 - W_3$) is equal to the amount of water transpired by the leaf (*ii*). The difference between (*i*) and (*ii*) gives the amount of water retained by the leaf.

Discussion: The above experiment clearly demonstrates that the amount of water absorbed by a twig or a leaf is not fully transpired under normal environmental conditions. Some amount of water is retained by it for its metabolic functions. This is a useful method for simultaneous determination of water absorption, retention and transpiration when the controlling factor acts chiefly through the leaves.

5

N.B. Instead of pouring oil in the conical flask before introducing the leaf into it, a vial containing oil may be weighed along with the flask and water. The leaf is then introduced and some amount of oil is poured from the vial on the surface of water. The vial should be always weighed in every step of weighing. This method will avoid the possibility of adherence of oil to the cut end of the petiole.

13. Demonstration of suction force due to transpiration or transpiration pull

EXPERIMENT: A narrow-bored glass tube is taken and a suitable leafy twig (cut under water) is inserted into it with the help of a rubber tubing and the connection is made air-tight with sealing wax. The tube is completely filled with water and holding it with the thumb, the lower end is carefully dipped below the surface of mercury contained in a small beaker. The whole arrangement is vertically clamped with the help of a stand and is kept in a well ventilated lighted place for transpiration. The rise in mercury column is noted at an interval of 15 minutes and final height is recorded after 1 hour.

RESULTS: The suction force due to transpiration or the weight of the transpiration pull is calculated by the formula $\pi r^2 \times h \times p \times g$ dynes per cm^2, where πr^2 is the area of the bore of the glass tube, h is the increase in the height of mercury column in the glass tube at a given time, p is the specific density of mercury (13·6) and g is the acceleration due to gravity (981 dynes per $sec.^2$).

DISCUSSION: The rise of mercury within the glass tube indicates the suction or the transpiration pull exerted on the water in the xylem vessels which is transmitted to the water column in the glass tube and ultimately to the mercury level in the beaker causing the mercury to rise in the tube.

N.B. This experiment can be further modified by taking into consideration:

(*i*) Stomatal and cuticular transpiration by smearing either surface with vaseline, (*ii*) the variation of the leaf area either by removing some leaves or smearing vaseline on both the surfaces of some leaves, (*iii*) herbaceous and woody twigs having comparable leaf area, and (*iv*) different environmental factors.

14. Determination of stomata-bearing surface of a leaf without using a microscope

EXPERIMENT: Four dorsiventral leaves of almost similar size and age are selected and vaseline is applied as follows:

(*i*) vaseline is smeared on both the surfaces of the first leaf,

(*ii*) vaseline is smeared on the upper surface of the second leaf,
(*iii*) vaseline is smeared on the lower surface of the third leaf, and
(*iv*) the fourth leaf is kept as such.

All the leaves are then weighed separately and kept hanging from threads which are tied to stands in an open space for a few hours. Now, each leaf is weighed again.

RESULTS: In case of the first leaf no wilting is observed and there is no loss in weight also. The second leaf wilts and loses weight. The third leaf, however, shows least wilting and minimum loss in weight. On the other hand the fourth leaf wilts considerably losing maximum weight.

DISCUSSION: Leaves lose water depending on the rate of transpiration. It is clear from the results that smearing of the upper surface of the leaves affects the rate of transpiration least. While smearing of the leaves on the lower surface considerably reduces the rate of transpiration, the upper surface in case of dorsiventral leaf is responsible for cuticular transpiration which is insignificant compared to stomatal transpiration through the lower surface. Hence, stomata are present on the lower surface in case of dorsiventral leaf.

15. Determination of the effects of environmental factors on the rate of transpiration

(*a*) EFFECT OF HUMIDITY:

EXPERIMENT: Conical flask oil water-leaf arrangement is to be used for this experiment. Two such arrangements having identical leaves are made. One such arrangement is weighed and kept in laboratory condition having relatively low relative humidity. The second weighed flask with the leaf is kept under a bell jar which is previously made humid by sprinkling water on the inner wall of the jar and by placing petridish containing steaming water which ensures almost saturated vapour pressure condition. The bell jar is placed on a glass plate and the contact is made air-tight with vaseline. After 2 hours both the flasks are weighed and areas of leaves are determined.

RESULTS: The rate of transpiration in each case is calculated and expressed as gm per sq. cm. per hour.

DISCUSSION: The internal atmosphere of the leaf is considered to be saturated or nearly so. The external atmosphere, on the other hand, is usually in an unsaturated condition. The vapour pressure gradient exists, therefore, between internal and external atmospheres and water vapour diffuses through the stoma from the region of high vapour pressure to the region of low vapour pressure. The lower the vapour pressure gradient, the less is the rate of transpiration at a constant temperature. If the vapour pressures of internal and external atmospheres are the same, no transpiration will take place. From the above experiment it is clear that the rate of transpiration under humid condition (high vapour pressure) is much less than the control.

(*b*) EFFECT OF WIND VELOCITY:

EXPERIMENT: Conical flask oil water-leaf arrangement is to be used for this experiment. Two such arrangements having identical leaves are made. The weighed control flask with the leaf is kept under laboratory condition in still air. The second weighed flask with the leaf is placed under a fan at a considerable distance from the first set. The leaf may be tied to the flask with thread to avoid its displacement. The experiment is continued for 2 hours. The flasks with leaves are reweighed. The areas of the leaves are determined.

RESULTS: Results are calculated and expressed as gm per sq. cm. per hour.

DISCUSSION: Air movements tend to hasten transpiration because it lowers the humidity of the external atmosphere of the leaf and thus make greater difference between the vapour pressure inside and outside of the leaf by mechanically removing the accumulated moisture. Whether the abrupt change in vapour pressure, which is one of the controlling factors of transpiration, is due to wind velocity is not known. The results of this experiment show that the rate of transpiration is much higher under air current than that under still air.

N.B. The wind velocity may be controlled by regulating the speed of the fan with the help of its regulator and the rate of transpiration under each air current may be compared.

(*c*) EFFECT OF LIGHT:

EXPERIMENT: Conical flask oil water-leaf arrangement can be conveniently used for this experiment. Three such arrangements having identical leaves are made. One weighed conical flask with leaf is kept under direct sunlight. The second weighed conical flask is kept under shade and the third weighed conical flask is kept in a dark room. The experiment is continued for 2 hours. The flasks with the leaves are reweighed and areas of the leaves are determined.

RESULTS: Results are calculated and expressed as gm per sq. cm. per hour.

DISCUSSION: Light and especially direct sunlight as contrast to shade or darkness, has a marked effect in bringing about an increase in transpiration.

This in most cases is chiefly due to the fact that much of the radiation is absorbed by the leaves. The temperature of the leaves in direct sunlight is, therefore, always higher than the surrounding temperature of the atmosphere. The second factor that might be of some importance is the opening of the stomata in the light. The results show that the rate of transpiration is maximum under direct sunlight and minimum under darkness.

N.B. Effects of different intensities of light may also be studied. Temperature is an important factor which controls the rate of transpiration. Effect of temperature may be studied by the conical flask arrangement keeping the flask in different temperatures.

16. Quantitative determination of transpiration under conditions of different experimental errors generally encountered

EXPERIMENT: During different transpiration experiments the following experimental errors are generally come across:

(*i*) The twig or petiole of the leaf not cut under water.
(*ii*) Adherence of vaseline to the surface of leaf through which transpiration is to occur.
(*iii*) Taking of aged or yellowing leaves as experimental materials for normal transpiration experiments.
(*iv*) Taking of leaves of different ages for similar experiments.
(*v*) Keeping up of experiments under closed environment.
(*vi*) Adherence of oil at the cut end of the petiole.

The same conical flask-oil-water-leaf-arrangement may be conveniently used in performing these experiments. Six such arrangements are made and each arrangement involves one of the above errors. Leaves from a same species of plant should be used for better comparison of results. A separate set-up is also made in which a healthy medium-aged leaf is taken avoiding all the above errors. All the flasks with leaves are separately weighed and kept in places which will suit the experiment most. The experiments are continued for about 2 hours. The flasks with the leaves are reweighed. The area of the leaves is measured.

RESULTS: Results are determined and expressed as gms. per sq. cm. per hour.

DISCUSSION: The petiole of the twig is cut under water in order to maintain the continuity of water column within the xylem vessels. If this continuity is broken air bubble gets entry and the transpiration pull stops. Vaseline is generally applied to the leaf surface to check transpiration by avoiding contact of the stomatal pore with the atmosphere. Thus, if vaseline adheres to a surface from which transpiration is to be determined, the transpiration is greatly affected. The rate of transpiration in case of aged or yellowed leaf is less because of its slow metabolic activity. Similarly different aged leaves affect transpiration in comparative studies owing to difference in metabolic activities as well as stomatal frequency. The rate of transpiration in a closed environment is affected by less light intensity and higher humidity. Adherence of oil at the cut end of the petiole hampers the entry of water through the conducting tissues. All these experiments show that precuations must be taken in performing experiments on transpiration to minimise the errors in results.

17. Demonstration of guttation

EXPERIMENT: Phenomenon of guttation can be seen in the early morning in plants such as *Bryophyllum*, *Tropeolum* (garden nasturtium), grasses, etc. especially after one humid night.

A small potted plant of tomato which has been previously well watered is taken and kept under a bell jar and sealed from all sides. The bell jar may be partially evacuated with the help of an aspirator fitted to the outlet of the bell jar. It is then placed in a warm room for some time.

OBSERVATION: Drops of guttation fluid oozing out at the margin of the leaves at its vein ending, are observed (Figure 17).

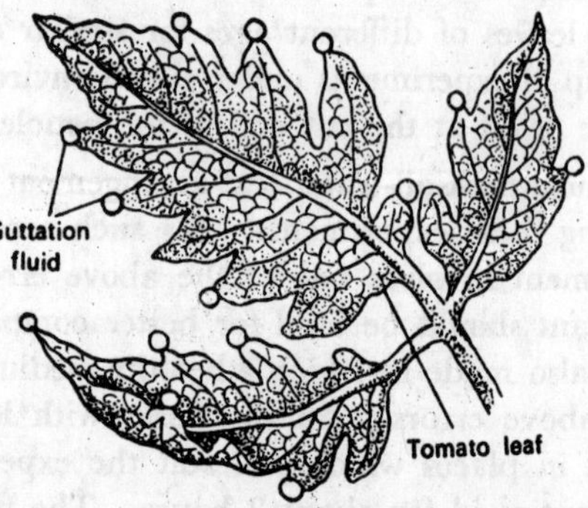

Fig. 17.

INFERENCE: Guttation is the exudation of liquid containing water from the margins of leaves of a real shoot. This occurs when the roots of such plants absorb water actively but transpiration is checked or lowered due to high relative humidity in atmosphere developing thereby a high root pressure. This produces a positive upward force in the xylem sap and causes exudation of liquid or guttation.

CHAPTER VII

MINERAL NUTRITION

Numerous researches have revealed that certain elements are necessary for the healthy growth of plants. They are termed as *essential elements* since growth and developmental processes of plants cannot proceed without them. The role of carbon, hydrogen and oxygen in the synthesis of carbohydrate, fat and related compounds is important in the life of the plant. In addition to these three elements, plants require a variety of elements which are usually grouped together under the class "the mineral elements" or "the mineral nutrients" Twelve mineral elements are essential for plant growth which are generally taken up by roots from the soil in the form of ions. According to their quantitative requirements, they are divided as *major elements* or *macro-nutrients*, as for example, nitrogen, phosphorus, sulphur, calcium, potassium and magnesium, and *minor elements* or *micro-nutrients* (also called *trace elements*), as for example iron, manganese, copper, zinc, molybdenum and boron (cobalt, silicon, sodium, vanadium, etc. are also included under micro-elements).

A deficiency of the above elements produces their characteristic symptoms, specific for the particular element involved. A study of these symptoms can greatly help in deciding the fertiliser requirements of the plants. By nutrient culture (*sand* or *water culture*) techniques, it is possible to study the deficiency symptoms of essential elements without much difficulty.

The analytical determination of the amount of a particular element present in the plant or in the soil in available form can be used to diagnose the quantitative requirement of the element for the normal growth of the plant. Plants suffering from the deficiency of a particular element will show its sub-optimal content in the plant body. Leaves are considered to be the best indicator and hence the "foliar diagnosis" has been widely employed.

1. Preparation of different culture solutions

Several early methods of studying plant nutrition are still found to be useful today. The analysis of plant ash and the use of liquid and sand cultures are techniques used for the study of plant nutrition in laboratories throughout the world. These methods, however, have been improved much.

A. Solution cultures:

This technique is useful because (*i*) it provides an excellent means for controlling the quantity and relative proportions of mineral salts given to a plant in any experiment, (*ii*) of the good solvent characteristics of water, and (*iii*) of the relative ease with which water can be freed from contaminating influences. Some of the sources of contamination, which is the main drawback of this technique, are the rooting medium, reagents used, containers, water, cutting implements, seeds and the dust in the surrounding atmosphere. Though it is impossible to eliminate contaminating influences, yet they can be kept to a minimum.

The first step is to prepare a stock solution from inorganic salts containing the necessary elements for normal plant growth. Once the stock solutions are prepared, nutrient solutions may be prepared by simply adding in the correct proportions the necessary inorganic salts from the stock solutions. For ready use some of the satisfactory formulae for nutrient culture solutions are given below. These solutions may be directly used for the experimental purpose.

(*i*) *KNOP'S SOLUTION (1860):

Salts	gm/litre
$Ca(NO_3)_2$, $4H_2O$	0·8
KNO_3	0·2
KH_2PO_4	0·2
$MgSO_4$, $7H_2O$	0·2
$FePO_4$	a trace

(*ii*) PFEFFER'S SOLUTION (1900):

Salts	gm/litre
$Ca(NO_3)_2$	0·8
KNO_3	0·2
$MgSO_4$	0·2
KH_2PO_4	0·2
KCl	0·2
$FeCl_3$	a trace

(*iii*) *Hoagland's solution (1920):

Salts	gm/litre Solution–1	gm/litre Solution–2
$Ca(NO_3)_2$, $4H_2O$	1·18	0·95
KNO_3	0·51	0·61
$MgSO_4$, $7H_2O$	0·49	0·49
KH_2PO_4 (for soln. 1)	0·14	—
$(NH_4)H_2PO_4$ (for soln. 2)	—	0·12
Ferric tartrate	0·005	0·005

(Either solution–1 or solution–2 may be used).

*It is advisable to add the trace elements B, Mn, Cu, Zn and probably Mo to the nutrient solutions given by Knop and Hoagland. These are supplied by adding 1 ml of Haas and Reed's so-called "a–z solution" per litre of nutrient solution. The "a–z solution" contains 8 elements which (*iv*) Arnon and Hoagland's solution (1940):

Salts	gm/litre	mg/litre
KNO_3	1·02	—
$Ca(NO_3)_2$	0·492	—
$(NH_4)\ H_2PO_4$	0·23	—
$MgSO_4$, $7H_2O$	0·49	—
H_3BO_3	—	2·86
$MnCl_2$, $4H_2O$	—	1·81
$CuSO_4$, $5H_2O$	—	0·08
$ZnSO_4$, $7H_2O$	—	0·22
H_2MoO_4, H_2O	—	0·09
$FeSO_4$, $7H_2O$ (0·5%)	—	0·60 ml/litre (three times weekly)
Tartaric acid (0·4%)	—	Same as above

have not been demonstrated to be essential for higher plants but omits Mo which is probably present as a contaminating element. Hass and Reed's "a–z solution" is as follows:

Salts	gm/litre
H_3BO_3	0·60
$MnCl_2$, $4H_2O$	0·40
$ZnSO_4$	0·05
$CuSO_4$, $5H_2O$	0·05
$Al_2(SO_4)_3$	0·05
KI	0·03
KBr	0·03
$Co(NO_3)_2$, $6H_2O$	0·05
LiCl	0·03
TiO_2	0·03
$SnCl_2$, $2H_2O$	0·03
$NiSO_4$, $6H_2O$	0·05

The roots of the plant to be grown are submerged in the nutrient solution, the stem projecting through an opening of the lid of the container. To give a more rigid system, the stem is generally held stationary in this opening by cotton padding. The container is to be covered in order to eliminate, as much as possible, the contamination caused by atmospheric dust. For good root growth and mineral salt absorption it is often necessary to provide some means of aeration.

(*v*) HEWITT'S SOLUTION (1963):

Salts	gm/litre or	mM/litre
KNO_3	0·505	5·00
$Ca(NO_3)_2$	0·820	5·00
NaH_2PO_4, $2H_2O$	0·208	1·33
$MgSO_4$, $7H_2O$	0·369	3·00
Ferric citrate	0·0245	0·10
$MnSO_4$	0·002230	0·01
$CuSO_4$, $5H_2O$	0·000240	0·001
$ZnSO_4$, $7H_2O$	0·000296	0·001
H_3BO_3	0·001860	0·033
$(NH_4)_6 Mo_7O_{24}$, $4H_2O$	0·000035	0·0002
$CoSO_4$, $7H_2O$	0·000028	0·0001
NaCl	0·005850	0·1

(Dilution technique should be used for the preparation of very low concentration of solution.)

B. SAND CULTURES:

Solid media, such as sand or crushed quartz or *vermiculite* (any of a number of micaceous minerals that are hydrous silicates derived usually from alteration of mica whose granules expand greatly at high temperatures to give a light-weight highly water absorbent material) are generally easier to work with than with a liquid medium. Today it is possible to get highly purified silica sand or crushed quartz which is very low in trace elements. One important aspect of sand culture is that the roots are growing in a natural medium and no means of support needs to be provided. Nutrient solutions are added to the solid culture by three different ways:

(*i*) SLOP CULTURE: pouring over the surface,
(*ii*) DRIP CULTURE: allowing to fall in drops over the surface,
and (*iii*) SUB-IRRIGATION: forcing solution up from the bottom of the container.

In all the three systems, the added nutrient solutions drain out through an opening at the bottom of the container. In sub-irrigation, the solution is collected in a reservoir and used repeatedly.

2. Determination of mineral deficiency by foliar diagnosis

The effects of deficiency of various essential mineral elements can be demonstrated by means of sand culture or water culture techniques. But the water culture technique is cumbersome for practical purposes because of the problem of aerating the culture solution every day and the possibility of frequent contamination. The sand culture technique is, therefore, more suitable where artificial aeration is not necessary. The culture solutions should be replaced daily to avoid danger from evaporation and contamination. The following stock solution will be required for this experiment. The molecular weight of each salt is given within parentheses.

Salt	Molarity (M)		
		Iron solution	
$CaCl_2$ (111)	1	$FeCl_3$	5 gm
$Ca(NO_3)_2, 4H_2O$ (196)	1	Tartaric acid	5 gm
$MgCl_2, 6H_2O$(203·24)	1	Distilled water	1 litre
$Mg(NO_3)_2, 6H_2O$ (256·32)	1	*Trace element solution*	
$MgSO_4, 7H_2O$ (246·38)	1	H_3BO_3	2·50 gm
KCl (74·56)	1	$MnCl_2, 4H_2O$	1·50 gm
KNO_3 (101·10)	1	$ZnCl_2$	0·10 gm
KH_2PO_4 (136·08)	1	$CuCl_2, 2H_2O$	0·05 gm
$NaNO_3$ (85·00)	1	MoO_3	0·05 gm
NaH_2PO_4, H_2O (137·98)	1	Distilled water	1 litre
Na_2SO_4 (142·06)	1		

The solutions are prepared for studying the effects of various elements according to the following table. The number of millilitres of each one molar single salt solution needed for one litre of the designated solutions is given below.

TABLE 2. Preparation of solutions for studying abnormal symptoms shown by plants due to deficiency of elements.

Salt solution	Quantity (ml)	Complete soln.	Salts to be added omitting the elements						
			K	Ca	Mg	N	P	S	Fe
KNO_3	2	+	−	+	+	−	+	+	+
KH_2PO_4	2	+	−	+	+	+	−	+	+
$Ca(NO_3)_2$	3	+	+	−	+	−	+	+	+
$MgSO_4$	2	+	+	−	−	+	+	−	+
KCl	2	−	−	−	−	+	+	−	−
$CaCl_2$	3	−	−	−	−	+	−	−	−
$(Na)_2SO_4$	2	−	−	+	+	−	−	−	−
NaH_2PO_4	2	−	+	−	−	−	−	−	−
$NaNO_3$	2	−	+	−	−	−	−	−	−
$Mg(NO_3)_2$	3	−	−	+	−	−	−	−	−
$MgCl_2$	2	−	−	−	−	−	−	+	−
Iron soln.	1	+	+	+	+	+	+	+	−
Trace element soln.	1	+	+	+	+	+	+	+	+

+ indicates presence and − indicates absence of a particular salt solution.

The required volumes of stock solutions obtained from the Table 2 are added in serial order to about 500 ml distilled water and the volume is made up to one litre. If the stock solutions are mixed together before dilution, precipitate will appear which can be dissolved only with difficulty.

Seeds are allowed to germinate aseptically in sand pots using distilled water. After the seedlings have grown to a uniform size, the distilled water is replaced with culture solutions. It will not be necessary to change the culture solution during the first two weeks, but thereafter at intervals of a week, the culture solution (excepting the trace element solutions) in each pot should be replaced with fresh solution. The sand may be kept moistened with distilled water.

OBSERVATION: The deficiency symptoms are observed according to the key given below.

Abnormal symptoms	Deficient element
A. PLANTS NOT PRIMARILY CHLOROTIC	
I. Necrosis insignificant	
Young leaves may appear abnormally dark green, but stems and leaves usually highly pigmented with purplish red, especially near the ends of shoots. Vegetative growth less than normal	Phosphorus
II. Necrosis pronounced	
(*i*) Plants usually not stunted, but leaves, especially the older ones, scorched and dying at tips and outer margins. Leaf margins often crinkled and curled	Potassium
(*ii*) Tips and bases of young leaves distorted, death of terminal buds. Young leaves of terminal buds at first hooked, later dying back at tips and margins, margins become dissected or broken. Stock finally dies ...	Calcium
B. PLANTS DEFINITELY CHLOROTIC	
I. Affected regions more or less uniformly yellowish; mottling (spotting) not conspicuous	
(*i*) Plants stunted, woody. Leaves, particularly the older ones, yellowish and drying up.	Nitrogen

(*ii*) Plants not particularly stunted. Leaves pale green, veins often somewhat lighter in colour than interveinal tissue. ... Sulphur

II. **Affected regions distinctly mottled**

(*i*) CHLOROSIS MOST APPARENT IN YOUNGER TISSUES
Younger leaves yellowish or even white, midribs and main veins often green, producing mottled appearance. Older leaves often uniformly chlorotic. Necrotic areas usually absent. ... ... Iron

(*ii*) CHLOROSIS MOST APPARENT IN OLDER TISSUES
Older leaves prominently mottled. Interveinal tissue pale yellow to white, midribs and main veins green. ... ... Magnesium

INFERENCE: The essential major elements are necessary for normal growth and physiological function of plants. Trace elements function in plants as constituent of specific enzymes and coenzymes. Moreover, some trace elements are also required in the biological electron transport system. The deficiency of a particular essential element results in the derangement of the complicated mechanism of the vital activities of plants.

3. **Influence of aeration of culture solutions on growth of plants**

EXPERIMENT: Equal quantities of culture solutions (Hoagland's) are taken in two wide-mouthed bottles of three litre capacity, according to the Expt. 1(a) and their pH is recorded. In one set-up aeration arrangement by bubbling through a rubber tubing is attached and the other is lacking such arrangement. Some freshly grown pea seedlings are transferred to each bottle. The bottle is aerated 6 hours every day till the end of the experiment. The growth data (average length of shoot and root, fresh and dry weights) are taken at an interval of 3 days. The pH of the solutions of the aerated and non-aerated bottles is recorded after the experiment.

RESULTS: Percentage increase in length of shoot and root and fresh and dry weights are recorded and graphically plotted.

DISCUSSION: The aeration of the solution is one of the most important aspects of solution culture techniques. In order to carry on the processes of growth and absorption of nutrients and water, it is essential that the living cells of the root have a supply of dissolved free oxygen for respiration. Another factor involved is the number of plants per culture bottle. When relatively few plants are present, the roots develop better owing to less competetion for oxygen. There is also a slower depletion of essential elements and a slower change in pH. Growth of plants, particularly in earlier stages, is usually better when air is forced to the culture solution mechanically, regardless of the type of container or composition of the solution.

4. Detection of nutrient deficiency occurring under field conditions

There are perhaps 4 general methods which are used for ascertaining the soil condition with respect to immediate plant nutrient requirements. They are, (*i*) plant diagnosis (already described), (*ii*) biological soil test (described below), (*iii*) chemical soil test, and (*iv*) chemical test for plant material. The last two chemical tests are beyond the scope of the present book and hence omitted.

BIOLOGICAL SOIL TEST:

A. ASPERGILLUS NIGER METHOD for determining Mg, Cu, Fe, Mn, Mo and Zn in soils and plant materials (Nicholas, 1945).

PRINCIPLE OF THE METHOD: This method of bioassay of mineral elements is based on the fact that *Aspergillus niger* requires specifically a number of mineral elements for its growth. These include the macronutrients —N, P, Ca, Mg, K and S and the micronutrients—Cu, Fe, Mn, Mo and Zn. When any one of these nutrients is omitted from the culture medium, the growth of the fungus is reduced much, as may be shown by dry weight yields of the felts and by the intensity and type of sporulation. A progressive increase in any one of these essential nutrients from deficiency to sufficiency level, provided others are present in optimum amounts, results in significant and quantitative increase in growth. A standard growth series may thus be prepared for each of the essential nutrients listed above.

A bioassay of test material for any given nutrient is made by adding a known amount of the material to a culture solution containing all the nutrients except that to be determined. The growth of the fungus under these conditions will depend on the amount of the test nutrient derived from the material added. The growth response is determined by comparison with a standard series for the elements prepared at the same time.

EXPERIMENTAL PROCEDURE:

(*i*) TYPE OF GLASSWARE: Hard glass 500 ml Erlenmeyer's flask and 150 ml beakers thoroughly cleaned by soaking in H_2SO_4-dichromate mixture or in dilute HNO_3 followed by washing with deionised water.

A simple test to detect the presence of metals on glass ware is to rinse with a 0·01 per cent solution of dithizone in chloroform. The presence of metal is shown by a change of colour from reddish brown of dithizone to red of the metal-dithizonate.

(*ii*) Purification of water: Good grade deionised water, checked by means of a conductivity bridge, is suitable for dispensing media.

(*iii*) Composition of basal culture medium:

Macronutrients (*gm*)		*Micronutrients* (*mg*)		*Dextrose* (*gm*)
KNO_3	5·0	$FeCl_3, 6H_2O$	20·0	50·0
KH_2PO_4	2·5	$ZnSO_4, 7H_2O$	20·0	
$MgSO_4, 7H_2O$	1·0	$CuSO_4, 5H_2O$	1·0	
$Ca(NO_3)_2$	0·5	$MnSO_4, 4H_2O$	3·0	
		$Na_2MoO_4, 2H_2O$	1·0	

Deionised water is added to one litre giving a solution of pH 3·8. For each culture 50 ml of solution is dispensed in a 500 ml Erlenmeyer's flask.

(*iv*) Purification of the medium:

Refer—"The diagnosis of mineral deficiencies in plants by visual symptoms" by T. Wallace, 1961.

(*v*) Choice of strain of fungus: *Aspergillus niger* M. has been found to be the most suitable strain.

(*vi*) Preparation of spore inoculum: The fungus is subcultured weekly on nutrient agar slopes by spore transfer technique. This is cultured at 30°C for four days. The composition of nutrient agar medium is as follows:

KH_2PO_4	2·4	gm
$MgSO_4, 7H_2O$	1·2	gm
Peptone	1·2	gm
Maltose	12·5	gm
Malt extract	12·5	gm
Agar	25	gm
Water	1	litre.

This medium is autoclaved at a pressure of 10 lb per square inch for 10 minutes and then dispensed in culture tubes and plugged with non-absorbent cotton wool and stored in a refrigerator at 4°C.

(*vii*) Transfer: The spores are transferred into 10 ml sterile deionised water and 0·1 ml of the inoculum is used for each 50 ml culture solution.

(*viii*) Preparation of standard series for the test metals: The sterile culture solution is transferred to clean 500 ml Erlenmeyer's flask in 50 ml lots. Necks of the flasks are covered with 50 ml inverted beakers. The fungus is grown at 30°C. for four days. The standard series may be assessed visually by taking into account the thickness of the mycelial growth, intensity of sporulation and quantitatively by dry weight yield of the felts. A standard series for a particular metal (different concentrations of the element) is to be prepared when an assay is to be made on test materials.

In the absence of Fe, Zn and Mo, thin mucoid mycelium devoid of spores are formed yielding between 20 and 50 mg dry weight compared with 900 mg for a normal mycelium grown in complete culture. The reduction in yield of felts grown without Cu or Mn is less striking than those for deficiencies of Fe, Zn and Mo although visually there is a marked difference.

PROCEDURE FOR ASSAY:

(*a*) SOILS: Soils are air-dried and passed through a sieve of 2 mm mesh and following amounts of soil are used for assay: 1 and 0·5 gm for Mg and 0·5 and 0·25 gm for Cu, Mn, Mo and Zn. The interference of micro-organisms in the soil may be overcome by adding 1 ml of 85 per cent redistilled ethanol per 0·5 gm of soil, removing the alcohol completely on a water bath. The fungus grows well between pH 1·8 and 8. For the assay of manganese in the soil, the best pH value is 7·5. The optimum pH for assay of Mo in soil is 2. The effect of *p*H on assay of Mg, Cu or Zn is less striking so that these can be done at the usual pH of the culture medium.

(*b*) PLANT MATERIALS: The trace metal contents of plant material may be determined directly or by ashing. The ashing may be done in platinum crucible at 400°C. after which the aliquots of the cooled residue dissolved in dilute HCl and transferred to the assay flask.

B. SOIL PLAQUE METHOD: The method is based upon the assumption that species of the nitrogen fixing bacterium *Azotobacter* commonly found in soils have nutrient requirements similar to those of higher plants so far as N, P, K are concerned. The extent of growth of *Azotobacter* on the surface of the moistened, variously treated soil samples is considered to indicate the soil requirements for these elements.

5. Demonstration of unequal absorption of cations and anions by plant

EXPERIMENT: Twenty-day-old seedlings of rice or wheat are suitable for this experiment. The following nutrient solutions are first prepared. The molecular weight of each salt is given within parentheses.

Solution–A	*Solution–B*
3·21 mM $(NH_4)_2SO_4$ (132)	4·84 mM KNO_3 (101·10)
0·12 mM $CaCl_2$ (111)	0·12 mM $CaCl_2$
1·39 mM KH_2PO_4 (136·08)	1·39 mM KH_2PO_4
3·92 mM $MgSO_4$, $7H_2O$ (246·38)	3·92 mM $MgSO_4$, $7H_2O$

Two lots of Solution–A are prepared and pH of one lot is adjusted to 4 and the other to 7 with 1% HCl and 1% KOH respectively. Solution–B is similarly adjusted to pH 5 and 7·5. The solutions are taken in separate Erlenmeyer's flasks and their levels are marked. Seedlings are taken out and roots are washed well without damaging them. Five such seedlings are placed in each flask. Bubbling of the solution may be done if possible. Distilled water should be added from time to time to make up the loss due to evaporation and uptake by plants.

OBSERVATION: pH of each of the solutions is recorded at two-day intervals with the help of a pH metre or Universal indicator paper.

DISCUSSION: Plants absorb cations and anions of different salts at different rates. It is difficult to measure these rates directly. But it can be demonstrated by an indirect method. Unequal uptake of cations and anions results in changes in the pH of the nutrient solution. If the cation of a salt is absorbed in excess of the anion, the pH will decrease (acidic) and if the anion is absorbed in excess, the pH of the medium will increase (alkaline).

6. **Quantitative determination of accumulation of chloride ions in plant cells**

EXPERIMENT: Weighed amount of *Hydrilla* plants is taken in a mortar and crushed with a little neutral sand and water. The extract is diluted to 50 ml with water after filtration. In order to remove chlorophyll pigment the solution may be treated with activated charcoal, if necessary. 10 ml of the extract is titrated against 0·02 N $AgNO_3$ solution (0·34 gm $AgNO_3$ per 100 ml solution) using 5% K_2CrO_4 solution as an indicator. As the end point reaches, a faint reddish brown colour of silver chromate ($AgCrO_4$) remains permanent when all AgCl has been precipitated.

Before the experiment, 0·02 N $AgNO_3$ is to be standardized against a standard NaCl solution (329·6 mg dry NaCl per 1000 ml solution) following the above procedure.

RESULTS: Calculations are made as follows–

Let the volume of 0·02 N $AgNO_3$ $= x$ ml
the volume of standard NaCl
for neutralisation $= y$ ml

Now, 1 ml of standard NaCl = 0·02 mg of chlorine.

Therefore, y ml of standard NaCl = 0·02 $\times y$ mg of chlorine.

Hence, x ml of 0·02 N $AgNO_3$ = 0·02 $\times y$ mg of chlorine.

Now 1 ml of 0·02 N $AgNO_3$ $= \dfrac{0{\cdot}02 \times y}{x}$ mg of chlorine.

If the volume of 0·02 N $AgNO_3$ to titrate 10 ml of plant extract be z, then the amount of chloride ion is $\frac{0{\cdot}02 \times y \times z}{x}$ mg per 10 ml of the sap.

The amount of chloride ion per gm plant material may be calculated.

DISCUSSION: In plant cell the vacuolar and plasma membranes are differentially permeable allowing water to pass freely but not the solutes. But under most conditions solutes (both electrolytes and non-electrolytes) may enter the cells passively. The electrolytes, on the other hand, may be actively accumulated within the plant cells. This active accumulation of ions requires energy which is supplied by respiratory breakdown of foods.

N.B. For determination of the chlorine per ml of the medium in which the plant grows, the medium (water) is taken and similarly titrated against 0·02 N $AgNO_3$. The difference of the chlorine content of the medium and that of the plant extract indicates the actual amount of chlorine accumulated by the plant.

CHAPTER VIII

PLANT IN RELATION TO pH

Some of the most fundamental physiological phenomena are markedly influenced by the concentration of hydrogen ions in the medium in which they occur. Thus hydrogen ion concentration in soil solution, in water culture solution, as well as in plant sap and protoplasm, is of great importance in the physiology of a plant. It affects solubility, availability and absorption of nutrients and innumerable others. It is, therefore, important to have an idea of the concentration of free hydrogen ions in a solution in which the organism grows. This is the pH scale. Hydrogen ion concentration may be stated in terms of its normality. But the expression of acidity in terms of normality is often awkward. A better shorthand expression was devised by the chemist Sorensen. Thus the pH is the common logarithm of the reciprocal of H-ion concentration expressed in terms of normality. Hence, pH is equal to $\log \frac{1}{[H^+]}$

The hydroxyl ion concentration of solution is the reciprocal of hydrogen ion concentration. This relationship may be given in the form of Mass law equilibrium equation as follows:

$$\frac{[H^+] \times [OH^-]}{[HOH]} = K = 10^{-14} \text{ at } 25^\circ\text{C. (ionic product for water).}$$

This constant is known as *dissociation constant*. When water is neutral, $[H^+] = [OH^-] = 10^{-7} =$ pH 7 (i.e., negative logarithm of 10^{-7}). With increase of acidity, H-ions increase and OH-ions decrease and pH is less than 7. When alkalinity increases, OH-ions increase but H-ions decrease and pH is more than 7. Physiologically active pH range is 3 to 9.

In most living systems the pH of the aqueous solution in which metabolic reactions proceed is maintained at remarkably constant value. This is provided through the presence of buffer system in plant. Buffers are chemicals that prevent marked changes in the pH of an aqueous system. The fact that certain organic acids, bases and salts dissociate to a relatively slight extent permits the formation of buffer system or solution. pK is called the negative logarithm of dissociation constant K. When the dissociation of acids and bases are equal, pK=pH. This happens when the solution is in its best buffering capacity. For example, the dissociation constant of acetic acid is $1{\cdot}8 \times 10^{-5}$, the negative logarithm is 4·7 which is equivalent to pK. A solution containing equal amount of acetic acid and sodium acetate will effectively hold the pH of the solution at $4{\cdot}7 \pm 1$ against the addition of acid or base. Below 3·7 and above 5·7, the pH will change rapidly.

1. Determination of total titratable acidity of the given plant sap

EXPERIMENT: 5 gm juicy hairs of lemon or orange or grape is taken in a mortar and crushed well with a pestle. This is then filtered using glass wool. The mother liquor thus obtained is measured and taken in a 100 ml volumetric flask. This is diluted to just 100 ml with distilled water (dilution factor $f=\frac{100}{\text{vol. of mother liquor}}$). This is then titrated against standard NaOH (N/10) using phenolphthalein as an indicator.

RESULTS: Let the volume of (N/10) NaOH be z ml.

Volume of diluted mother liquor be y ml.

Therefore, z ml of (N/10) NaOH is equivalent to y ml of x(N/10) diluted mother liquor, where x is the strength of diluted mother liquor.

Now, $V_1 \times S_1 = V_2 \times S_2$

$$z \times (N/10) = y \times x(N/10)$$

$\therefore x = z/y(N/10) = z/10y(N)$. Hence the strength of mother liquor $= \frac{z \times f}{10 \times y}$ (N), where f is the dilution factor.

DISCUSSION: Newly produced in metabolism, acids remain in free state and increase the titratable acidity (the free acids which can be titrated against an alkali) of the cell sap and tend to lower the pH. Subsequently, however, the titratable acidity may decrease without change in the amount of total acidity.

Organic acids like oxalic, tartaric and Krebs cycle intermediates, viz., malic, isocitric, citric, and other acids are produced and may accumulate in different plant parts, particularly in green fruits. Accumulation implies a rate of enzymic production exceeding that of consumption. The formation and consumption of oxalic and tartaric acids which are not Krebs cycle acids cannot be explained. There is a wide fluctuation in content of individual acids among different genera of plants.

2. Determination of pH of the cell sap of the plant material

EXPERIMENT: The original mother liquor of the previous experiment is taken and its pH is determined either by using universal pH indicator paper and comparing the coloured chart, or by the indicator prepared according to the following formula: Phenolphthalein — 0·1 gm, Methyl red — 0·2 gm, Dimethylaminoazobenzene — 0·3 gm, Thymol blue — 0·5 gm, Bromothymol blue — 0·4 gm, Absolute alcohol — 500 ml.

A drop of the mother liquor is taken on a spot plate, mixed with a drop of indicator and pH is determined by comparing the colour chart.

For standard comparison, 5 ml of sample and 2 drops of indicator solution give better results.

N.B. pH may also be determined more accurately with the help of a pH-metre.

3. Determination of degree of dissociation of cell sap

EXPERIMENT: If pH or total acidity of any acid or cell sap can be known, the degree of dissociation can be calculated as follows:

Let the strength of HCl = 0·1 (N), pH = 1·14 (strong acid) and strength of acetic acid = 0·1 (N), pH = 1·85 (weak acid).

Now, antilog of 1·14 = 14, therefore its reciprocal = 1/14 or 0·07 N is the hydrogen ion conc. in normality

and antilog of 1·85 = 71, therefore its reciprocal = 1/71 or 0·014 N is the hydrogen ion conc. in normality.

Hence, the degree of dissociation of 0·1 N HCl = (0·07/0·1) ×100 = 70%

and the degree of dissociation of 0·1 N CH_3COOH = (0·014/0·1) ×100 = 14%.

In case of cell sap, the pH and strength in terms of 0·1 N NaOH are to be determined first according to the experiments already described. Then the degree of dissociation is determined as above. The degree of dissociation increases with increase in dilution of acids or alkalis.

4. Determination of buffering capacity of plant sap

The efficiency of a buffer system (or its buffering capacity) depends upon how many moles of acid or base can be added before we get an appreciable change in the initial pH. In any case the buffering capacity depends on the actual amount of acid and conjugate base available for combining with the added base or acid.

The buffering capacity of cell sap, soil solution or any other liquid can be expressed by a number, the *buffer index*. A solution has a buffer index of unity when the addition of 1 gm equivalent of a strong acid or alkali shifts the pH of 1 litre of the solution through one unit. In general, it may be said that the amount of shift depends upon the nature and concentration of the buffer. Comparative values of the buffer capacity of different solutions may be obtained by determining with pH-meter or indicators, the number of gram equivalent of acid or alkali that must be added to equivalent amount of a solution to produce unit change in pH. Biological fluids are in general strongly buffered, so that their pH is maintained within narrow limits under physiological conditions.

EXPERIMENT: 25 ml of the plant juice is taken in a 100 ml flask. pH of the cell sap is determined with the help of a pH-meter or a universal pH indicator paper. 1 ml of 0·5 (N) HCl is added from a burette to the sap and the beaker is gently shaken. Its pH is again determined. The addition of 1 ml aliquot of acid is continued and pH is determined in each case until pH value of 2 is reached. To a second 25 ml of original juice or sap 0·5 (N) NaOH is added from a burette in 1 ml increments until a pH value of 12 is reached. The pH is determined each time after addition of each ml of 0·5 (N) NaOH.

RESULTS: The results of the titration are plotted on a graph paper in the form of two titration curves using a single co-ordinate axis taking pH increasing from 2 to 13 horizontally along the abscissa. At the midpoint of the ordinate zero is placed. Now equivalents of alkali added are placed above it and equivalents of acid added placed below it. The buffer capacity can be estimated from the slope of the titration curve. Sharp slope indicates a large pH change for small change in base or acid concentration, i.e., a poor buffer capacity and *vice-versa*. A model graph is given below for better understanding (Figure 18).

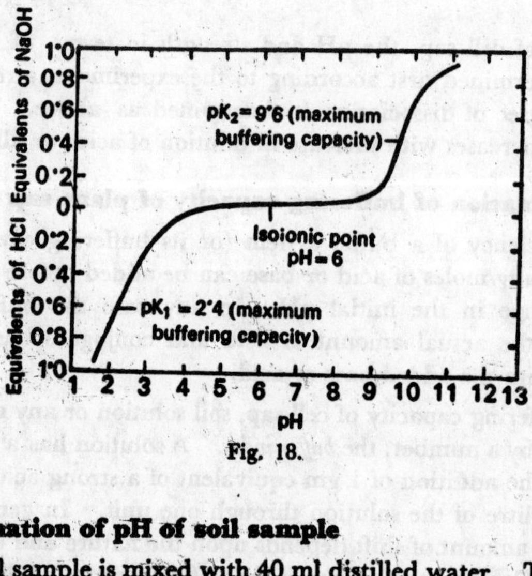

Fig. 18.

5. Determination of pH of soil sample

20 gm soil sample is mixed with 40 ml distilled water. The suspension is stirred at regular intervals for ½ hour. pH is then measured in a pH meter. The suspension is to be stirred well before taking reading of the sample.

6. **Determination of pK value of plant sap and the possible buffering acids present**

The pH at which the plant sap shows the maximum buffering capacity is equivalent to its pK value. If maximum buffering capacity is shown, say, at pH 4·74, then pK is equal to 4·74. The antilog of this value is equal to $1{\cdot}8 \times 10^{-5}$ and pK is, according to the definition, the negative logarithm of the dissociation constant K. Now, the dissociation constant is equal to $1{\cdot}8 \times 10^{-5}$ which is typical for acetic acid (pK = 4·74, 5 − 4·74 = 0·26, antilog of 0·26 = 1·82, hence its negative logarithm = $1{\cdot}82 \times 10^{-5}$ = K).

The dissociation constant can also be determined from the degree of dissociation (percentage).

$$K = \frac{[H^+] \times [A^-]}{HA}$$

$$[H^+] = \frac{\text{percentage of dissociation}}{100} \times C\ (N) = a$$

where C is the strength of acid in normality.

$$[A^-] = \frac{\text{percentage of dissociation}}{100} \times C\ (N) = a$$

$$HA = (1 - a) \times C\ (N)$$

$$K = \frac{a \times a}{(1 - a) \times C} = \frac{a^2}{(1 - a)\ C}.$$

Some of the typical dissociation constants which help in identifying the compounds are as follows:

Acetic acid = $1{\cdot}8 \times 10^{-5}$

Phosphoric acid = $1{\cdot}1 \times 10^{-2}$, 2×10^{-7} and $3{\cdot}6 \times 10^{-13}$

Carbonic acid = 3×10^{-7} and 6×10^{-11}

CHAPTER IX

TRANSLOCATION OF ORGANIC AND INORGANIC SOLUTES

Translocation is one of the important processes in the physiology of plants. There are two distinct types of translocation involving different routes and directions, one is the translocation of organic solutes downward through the phloem and the other inorganic solutes upward through the xylem. Organic translocation is the movement of food stuff within the plant from the site of synthesis to the site of utilisation. It appears to fulfil the functional requirements of the plant. The organic materials can move up or down, independent of the water stream, insulated from the cells of the plant, and in system capable of repairing breaks. The direction of flow correlates well with sites of utilisation, such as toward growing point, and roots, but there is as yet no satisfactory explanation of how the flow is steered through the plant. The accumulation of organic materials in fruits and storage organs must involve some mobilisation force which the mass flow theory assumes but does not explain. The capacity for mending breaks in the phloem tube system appears to be a function of the sieve plates and associated slime plugs, though the function of these structures in permitting flow is not thoroughly understood.

In majority of plants the inorganic materials absorbed by the roots are translocated upwards through the xylem along with the transpiration stream. The distinction between organic and inorganic translocation almost disappears when close comparisons are made. Many erstwhile inorganic nutrients apparently move in the translocation systems as organic forms or as chelates with organic materials. There is not even reasonable distinction between the kinetics of those that do move in the inorganic form and those that do not.

Three general types of factors regulating translocation may be recognised. (*i*) The movement of the xylem and phloem sap is a major factor, (*ii*) features of the solute (solubility, diffusion characteristics, partitioning properties with the cell wall or other non-moving components, and any tendency to be removed by surrounding tissues) can modify its ability to move with the solvent, and (*iii*) there is a metabolic involvement most evident in the directive activities of the plant apex, in the retranslocation from older leaves, and in the mobilising forces in general. Though metabolic activities may direct the destination of the nutrient, they do not seem to be involved in the actual flow.

1. Demonstration of upward translocation from germinating seeds

EXPERIMENT: About 100 seeds of pea or gram or *Vicia* seeds are soaked in distilled water. The seeds are divided into two lots. One lot of seeds is taken out at the stage when seed coats can be removed. The fresh weight, dry weight and ash weight of this lot are determined. The other lot is allowed to grow in dark. As soon as the primary leaves appear in this lot, the cotyledons are separated from a few seedlings. (*i*) The average dry weight of the cotyledons and the shoot, (*ii*) the average ash weight and (*iii*) the average loss or gain of the cotyledons and the shoots are separately determined.

RESULTS: (*i*) Percentage loss of dry weight and ash weight of the cotyledons and (*ii*) the corresponding percentage increase of dry and ash weights of the shoot are calculated from the data.

DISCUSSION: The cotyledons are the storehouse of growing embryo. As the embryo grows to a seedling, reserved food material is translocated to the seedling from cotyledons. So long the leaves are not formed, photosynthesis cannot take place and seedlings remain entirely dependent on the cotyledons for their nourishment and growth. In this experiment the seedlings are grown in dark in order to preclude the possibility of getting nourishment of the seedlings from the photosynthate. Decrease of the food materials from the cotyledons and concomitant increase in the shoot indicates that food materials are translocated upward to the shoot at the cost of cotyledons. Some amount of food materials is lost by way of respiration or other catabolic processes in both cotyledons and shoot which may be taken here as insignificant.

2. To show the downward translocation of food in a woody stem (or effect of ringing upon food movement)

EXPERIMENT: The experiment is to be performed with a woody plant in which the apical growth has ceased. It is best performed in growing season. One or more stems or branches are selected which have no side branch for 50 cm or more and which are several centimeters in diameter. Ring is made by removing the bark and phloem tissue approximately in the middle of the clear portion. The removed rings of bark should be about 1 cm wide and ringing should be done carefully so that the xylem is not damaged. The exposed surface of the xylem should be carefully scrapped with a sharp knife so that all traces of cambium are removed. The exposed surface is covered with paraffin wax.

After three weeks, final growth measurements are taken. Sections are cut 25 to 50 cm from the regions immediately above and below the ring and following data are taken: (*i*) Sections from just above and just below the ring are tested for starch with iodine solution and compared. (*ii*) The volume of 100 gm of tissue from below and above the ring is determined by displacement of water and compared. (*iii*) The percentage of dry matter in each tissue is determined and compared.

RESULTS: Results are tabulated or plotted to show the effect of ringing upon translocation of food material based on the above three indices.

DISCUSSION: Food materials are synthesized in leaves and translocated downwards through the phloem. Removal of phloem tissue hampers this downward translocation and accumulation of food materials above the ring occurs. This experiment indicates that organic solutes flow downward through the phloem into root and other organs, when the above three indices below and above the ring are compared.

3. Demonstration of translocation from leaves

EXPERIMENT: Several seedlings of kidney bean (*Phaseolus vulgaris*) are chosen on which the first pair of primary leaves is well developed. Three sets of plants are exposed to bright sunlight until the leaves show heavy starch accumulation on test with 1% iodine. One petiole of each pair of leaves is then treated by the following methods, leaving the second petiole and leaf intact as control. (*i*) One petiole is cut off and kept in water for comparison with control leaf; (*ii*) killing a portion of one petiole by a hot forceps (heating any other portion of the plant is avoided; the leaf may be kept in position with the help of a thread); and (*iii*) a portion of one petiole is anesthetised with chloroform or ether soaked in cotton. All the plants are kept in dark in a moist chamber at 20°C. for 24 hours. Chlorophyll is removed from the leaves with alcohol and a few drops of lactic acid and tested for starch with 1% iodine solution.

OBSERVATION: The untreated leaves show little or no response with iodine test compared to treated ones.

INFERENCE: Marked loss of starch from a leaf is taken as an evidence of downward translocation. Considerable retention of starch in a treated leaf indicates that starch has not been translocated due to interference in the translocatory path. The experiment thus shows that food material is translocated from leaves.

4. Demonstration of upward translocation of food in woody stem

EXPERIMENT: The experiment can be best performed in growing season. Four approximately uniform twigs on a woody plant are selected. Potted woody plants or even cut branches with their basal ends immersed in water can be used in this experiment. When tested with iodine they show considerable quantity of starch. Smooth-barked species with rather stout stems and with true terminal buds are most suitable. If possible twigs should be selected from plants which bear no side branches for a distance of about 40 cm back from the terminal bud. The selected branches are numbered. Twig number 1 is ringed about 5 cm, twig number 2 about 20 cm, and twig number 3 about 40 cm below the respective terminal buds. Twig number 4 is kept as control. During ringing, a strip of bark is removed 0·5 cm wide and the exposed wood is carefully scrapped to remove all traces of cambium avoiding any damage to the xylem. The exposed wood is coated with paraffin wax. Leaves and lateral buds are removed from the portions of the twigs above the rings as fast as they emerge from the stem. On the control stem all the leaves and lateral buds which start to develop about 40 cm below the original location of the terminal bud are removed. The twigs are observed from time to time as growth proceeds from the terminal bud, noting especially differences in the rate of longitudinal growth.

Results: After three weeks the stem elongation which has occurred from the terminal bud of each stem is measured and recorded in millimeters. Cross sections are cut from above and below the ring of each stem and tested for starch with iodine solution.

Discussion: The rate of growth above the ring in case of number 1, 2 and 3 twigs is very much checked as compared with control. The growth that has occurred in the ringed stems is only due to upward translocation of food materials through xylem. In case of control twig the upward translocation has taken place both through xylem and phloem. It is also evident from the experiment that the more is the distance of the ring from the terminal bud, the less is the rate of growth above the ring. The accumulation of starch is always maximum just below the ring.

N.B. From the above experiment a correlation with translocation and growth may be made. The influence of phloem upon translocation and growth may also be studied by removing different amounts of phloem tissue from a particular region on the stem.

5. To demonstrate the exudation from phloem tissue

Experiment: *Cucurbita* seedlings are grown under favourable conditions until they have attained a length of 30 cm. The stem of one of the plants is cut off with a sharp scalpel from 5 to 10 cm above the soil surface. The cut end of the excised portion of the stem is held in an inverted position and observed under a powerful hand lens or binocular microscope for exudation. Exudation may also be studied by puncturing a sharp needle through the bark to a sufficient depth to just reach the inner layer of phloem.

Observation: It is observed from what tissue (xylem or phloem?) the exudate comes. If exudate cannot be discerned clearly in the excised stem the first drop of exudate is blotted with a filter paper and the cut enc is re-examined.

INFERENCE: It is clear from the study that the sap comes out mainl through the phloem tissue in the form of droplets.

6. Demonstration of translocation of food into developing fruit

EXPERIMENT: Suitable species of fruit trees on which fruits of consider able size usually develop are selected. The experiment should be starte when the fruits on the tree are half-matured. At least ten fruits are tagge and numbered and their circumferences are carefully measured. Th fruits are isolated from the main phloem system of the plant by means c proper ringing. Rings may be made at the base of fruiting branch o both above or below its point of attachment on the stem. All precaution are followed in ringing the stem. The circumferences of an equal numbe of fruits from unringed branches are also measured to serve as controls From time to time both sets of fruits are measured and rates of growth i diameter of the fruits from the ringed and unringed branches are com pared.

RESULTS: The rates of growth in diameter of fruits in ringed and contrc sets are recorded and compared.

DISCUSSION: The developing fruits are the active centres of mobilisatio of carbohydrate from other regions of the plant. Since phloem is th principal path of translocation of food materials to the developing fruits removal of phloem tissue greatly hampers this transport of carbohydrate t the growing fruits. Hence the growth rate of the ringed fruits is muc less compared to controls.

7. Demonstration of upward translocation of mineral salts i woody stem

EXPERIMENT: This experiment can be performed with potted wood plant as in Expt. 4. Three sets of at least five comparable branches on th plant are selected. In one set all of the branches are ringed 5 cm below th base of each terminal bud. Rings are made about 0·5 cm wide, the ex posed wood is carefully scrapped to remove all traces of cambium and th exposed surface is covered with paraffin wax. In the second set all th branches are removed from the plant by making a sharp cut at the poin corresponding to that at which the branches of the first set were ringed, i.e. 5 cm below each terminal bud. This set is used as 'starting' control. In th third set all the branches are tagged at a point 5 cm below the termina bud. At the end of three weeks the branches of the girdled set and th tagged set which is to serve as 'end' control are removed by cutting then off at the point of girdle and at the point of tagging respectively. The ash content of the branches of the 'starting' control is determined at the beginning of the experiment and that of the 'end' control and of ringed branches at the end of the experiment, as follows. The sample of stems to constant weight is dried in an oven at about 80°C. and its dry weight is determined. Each sample is ground and mixed thoroughly. The ground dry tissue from each set of stem is weighed and heated to constant weight in a muffle furnace at about 600°C.

Results: The estimated ash contents are expressed as percentages of the dry weight and fresh weight of stems.

Discussion: The mineral from the soil solution is carried through the xylem in transpiration stream, though some amount of mineral is translocated through phloem. As minerals are mainly translocated through the xylem, the removal of phloem tissue from the ring will not debar the stem above the ring from the supply of mineral nutrients. When the ash content of the stems of three sets is compared, it becomes clear that percentage of ash content is maximum in case of the set where the branches were originally removed 5 cm below the terminal bud and minimum in case of the set where branches remained intact. This shows that translocation of mineral solutes takes place mainly through xylem.

8. Demonstration of the effects of inhibitors on the uptake, distribution and translocation of ^{32}P in plants

Experiment: One-month-old bean seedlings grown in sand culture (Ref. Chapter VII) may be suitably employed in this study. The plants are removed from the culture taking sufficient care not to injure the roots. Roots are then washed well to remove the adhering particles. These are then selected for treatment. Two such plants are taken and the roots are inserted in each of the six test tubes containing the following solutions (10 ml) and three test tubes may be bubbled for aeration.

1. 6 μc of ^{32}P/12 ml of 0·01 M Na_2HPO_4
2. 6 μc of ^{32}P/12 ml of 0·01 M Na_2HAsO_4 in 0·01 M Na_2HPO_4
3. 6 μc of ^{32}P/12 ml of 0·001 M NaN_3 in 0·01 M Na_2HPO_4

Now from each leaf, discs are prepared with the help of a corkborer at different intervals, dried under infra-red lamp and its radio-activity is measured in a Geiger-Muller Counter. After 1 to 2 hr. the plants are taken out, roots are washed well with carrier phosphate (0·01 M Na_2HPO_4) and the distribution of radio-activity is determined by autoradiography as follows. The plant is kept for sufficient time in contact with a suitable film in a dark room under some uniform pressure. **[Half-life of ^{32}P is to be taken** into consideration (14·3 days)]. After exposure the film is developed and the spots of radio-activity are determined. *Exposure time* $= \frac{10^7}{x}$ min. $+3$ days, where x is the count per minute (CPM) during experiment.

OBSERVATION: The uptake of ^{32}P by plants as affected by aeration, sodium arsenate and sodium azide is noted and data are tabulated.

DISCUSSION: The above experiment clearly demonstrates the translocation of radio-active phosphorus to different organs of the plant. The translocation of phosphorus is a function of time and distance from the roots as indicated by the radio-activity in different plant parts. Again metabolic inhibitors like azide or arsenate inhibit the rate of translocation indicating that it is regulated by the metabolism of living cells. Increased translocation with aeration suggests that respiratory energy is also involved in the translocation of phosphorus.

CHAPTER X

RESPIRATION AND FERMENTATION

Respiration is a catabolic process in which energy is released involving oxidation of organic compounds with the molecular oxygen which serves as an ultimate electron acceptor. During the process of photosynthesis light energy is converted to chemical energy and conserved in the bonds of complex organic molecules. The major portions of stored energy in plants may be found in the form of carbohydrates. The breaking of the C–C bonds of such compounds may release a considerable amount of energy for utilization by the plant. This bound chemical energy in organic food materials is released by oxidative reaction of respiration. However, the total amount of energy content within a compound is not released all at once but slowly, in stepwise series of reactions, controlled by enzymes. Respiration is characterised by oxidative degradation of complex organic compounds within all living cells, releasing energy, CO_2, H_2O and other simpler compounds from which synthesis of many other complex compounds can take place.

The summary reaction of complete breakdown of 1 gm molecule of hexose sugar with the liberation of energy is given below.

$$C_6H_{12}O_6 + 6O_2 \longrightarrow 6CO_2 + 6H_2O + 684 \text{ K. Cal.}$$

It means that one gram molecule hexose or 180 grams of it, on complete oxidation, yields 684 K. Cal. of energy, or 1 gram of hexose yields 3·8 K. Cal. of energy. Proteins and fats yield a larger amount of energy. Protein on complete oxidation yields about 5·7 K. Cal. per gram of protein oxidised, whereas fat yields on an average 9·1 K. Cal. per gram oxidised.

There are mainly two types of respiration, viz., (*i*) *aerobic respiration* in which complete oxidation of respiratory substrate takes place in presence of oxygen through EMP-linked TCA cycle resulting in end products of carbondioxide and water, and (ii) *anaerobic respiration* or *fermentation* in which respiration occurs in absence of oxygen when the respiratory substrate is partially oxidised to form some carbonic compounds like ethyl alcohol, acetic acid, lactic acid, etc., along with CO_2 and H_2O. The energy released in the anaerobic respiration (24 K. Cal. per gm mol.) is much less than that of aerobic respiration.

AEROBIC RESPIRATION

$$C_6H_{12}O_6 + 6O_2 \longrightarrow 6\,CO_2 + 6H_2O + 684 \text{ K. Cal.}$$

ANAEROBIC RESPIRATION OR FERMENTATION

In absence of O_2

$$C_6H_{12}O_6 \longrightarrow \underset{\text{[Ethyl alcohol]}}{2\,C_2H_5OH} + 2\,CO_2 + 24 \text{ K. Cal.}$$

Respiratory quotient or RQ is the ratio of volume of CO_2 evolved to the volume of O_2 used in respiration by a given mass of tissue in a given time interval.

$$RQ = \frac{\text{volume of } CO_2 \text{ evolved}}{\text{volume of } O_2 \text{ consumed}}.$$

The value of RQ of a tissue depends upon the nature of substance which acts as a respiratory substrate and is oxidised in respiration. If the RQ of a tissue is known it is possible to guess the nature of respiratory substrate. When carbohydrate is the respiratory substrate RQ is unity. $C_6H_{12}O_6 + 6\,O_2 \longrightarrow 6\,CO_2 + 6\,H_2O$, $RQ = \frac{6\,CO_2}{6\,O_2} = 1$, when glucose is completely oxidised.

When respiratory substrate is fat or protein, RQ is less than one. $2\,C_5H_8O_6 + 145O_2 \longrightarrow 102CO_2 + 98\,H_2O$, $RQ = \frac{102\,CO_2}{145\,O_2} = 0{\cdot}7$, when fat is completely oxidised. Here the proportion of O_2 to CO_2 is less than that in carbohydrate. Hence larger amount of O_2 will be required for oxidation of fats. In case of protein the value of RQ is less than one, because proteins are made up of amino acids, which require more oxygen for complete oxidation than carbohydrates. The RQ of proteins is generally 0·8.

When respiratory substrate is organic acids, the RQ is generally more than one, because less oxygen is required for their oxidation. For example, RQ of malic acid is 1·33, oxalic acid 4, tartaric acid 1·6 and so on.

$$C_4H_6O_5 + 3O_2 \longrightarrow 4\,CO_2 + 3H_2O$$

Malic acid

$RQ = \frac{4\,CO_2}{3\,O_2} = 1{\cdot}33$. The organic acids are rich in oxygen as compared with carbohydrate, little oxygen will be required for oxidation process and consequently the value of RQ is greater than one.

The value of RQ deviates from unity under the following conditions. (*i*) When substrates other than carbohydrates are used, (*ii*) when partial oxidation of substrate occurs, (*iii*) when O_2 absorbed is used in a process other than respiration, (*iv*) when CO_2 evolved is again absorbed in some processes, and (*v*) when CO_2 is evolved by a process other than respiration.

The rate of respiration may be influenced by several environmental factors, such as temperature, oxygen concentration, carbon dioxide concentration, concentration of inorganic salt in the culture medium, mechanical handling and wounding, etc.

Many types of aliphatic carboxylic acids like citric acid, malic acid, succinic acid, etc., are intermediate reactants in the sequence of reaction in the TCA-cycle and so they are found in considerable amounts inside the cell. They are thus directly involved in the respiration process. Certain plants belonging to families Crassulaceae, Cactaceae, Euphorbiaceae, etc., show synthesis of organic acids in abundant quantity during night. But these acids are again broken down during the day time producing CO_2 which is again photosynthetically fixed. This process of acid formation in dark and deacidification of these acids in light is termed *Crassulacean acid metabolism* (CAM).

1. Demonstration of aerobic respiration in plants

EXPERIMENT: Aerobic respiration in plants can be experimentally proved with the help of a simple apparatus like (*i*) respiroscope which consists essentially of a stout vertical tube which is bent into a bulb at the one end (Figure 19a), or (*ii*) with the help of a long-necked round-bottomed

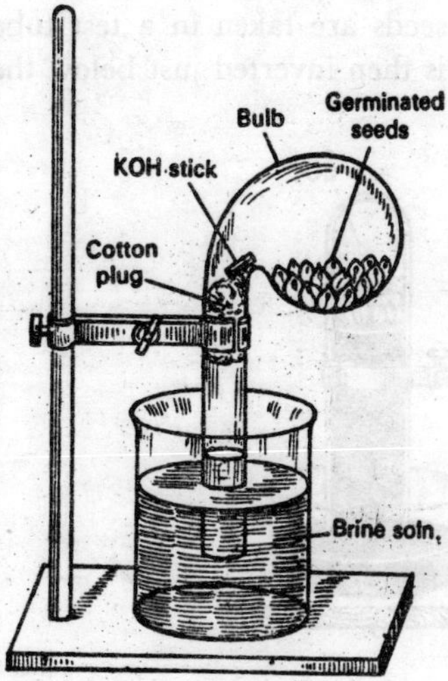

Fig. 19(a).

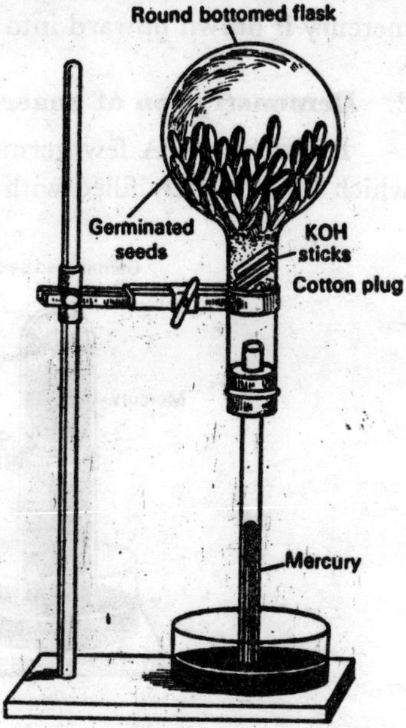

Fig. 19(b).

flask fitted with a centrally-bored cork at the mouth through which passes a glass tube (Figure 19b).

The respiroscope or the inverted flask is fixed vertically to a stand and a few germinating gram seeds or flower petals are placed in the bulb of the respiroscope or in the inverted flask plugged with cotton at the base of the neck. The vertical tube of the respiroscope or inverted flask is dipped just below the surface of water or mercury in a beaker. A few caustic potash (KOH) pellets are introduced in the bent portion of the respiroscope or in the long neck of the round bottom flask and kept in position with loosely held cotton wool. Care should be taken that respiratory materials and KOH pellets do not come in contact. Precautions should be taken that the free end of the tube does not touch the bottom of the water or mercury trough. Fittings must be air-tight to avoid any leakage.

OBSERVATION: The apparatus is allowed to stand for a few hours when it is seen that water or mercury has risen in the vertical tube of the apparatus proving the production of partial vacuum.

INFERENCE: Due to respiration of germinating seeds or flower petals CO_2 has been released which is at once absorbed by KOH pellets. Thus the partial vacuum produced by the absorption of O_2 by the respiring material could not be filled up by the released CO_2. Hence, water or mercury is drawn upward into the tube.

2. Demonstration of anaerobic respiration

EXPERIMENT: A few germinating gram seeds are taken in a test tube which is completely filled with mercury and is then inverted just below the

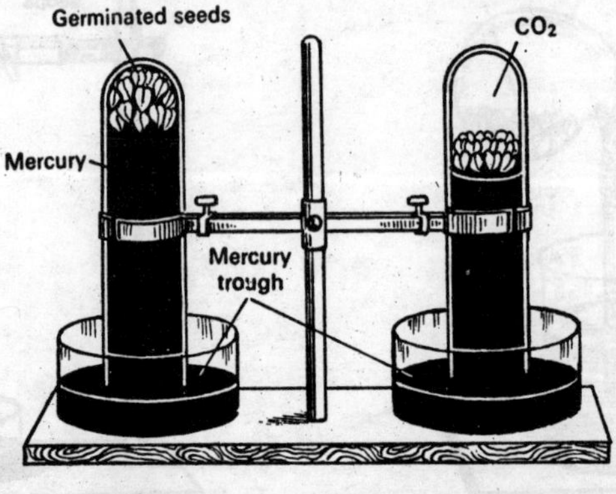

Fig. 20.

surface of mercury in a trough. It is then vertically held with a clamp and stand (Figure 20).

OBSERVATION: Observation from time to time reveals that a gas is formed within the test tube by the displacement of mercury in the test tube. A few KOH pellets are introduced through the open end of the test tube when mercury again rises filling the test tube.

INFERENCE: Here the respiration of germinating seeds takes place in complete absence of O_2 supply and the gas produced is CO_2 as evidenced by its absorption by KOH. This proves that anaerobic respiration has taken place.

3. Demonstration of alcoholic fermentation

EXPERIMENT: A fermentation tube or Kuhne's vessel (Figure 21) is filled with 10% sucrose solution and mixed with a small quantity of Baker's yeast or a few millilitres of suspension of yeast cells. The open end of the apparatus is plugged with cotton wool.

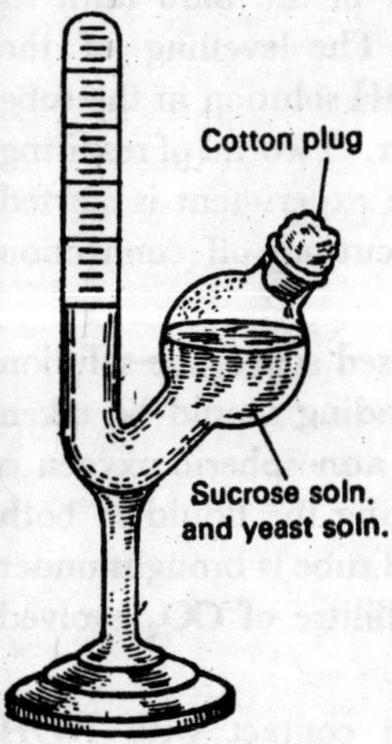

Fig. 21.

OBSERVATION: Occurrence of fermentation or anaerobic respiration and collection of CO_2 gas in the back arm of the Kuhne's tube are observed. When the cotton wool is taken off smell of alcohol may be perceived.

INFERENCE: In alcoholic fermentation the sugar is broken down to alcohol and carbon dioxide liberating certain amount of energy. The enzyme, 'zymase complex' present in the yeast brings about this reaction through a number of steps. Alcoholic fermentation takes place in absence of atmospheric oxygen. Hence it is an anaerobic process.

4. Determination of rate of respiration

EXPERIMENT: The rate of respiration can be measured with the help of Ganong's respirometer (Figure 22).

DESCRIPTION AND EXPERIMENT: The apparatus consists of three parts: (*i*) The bulb for the respiring material which ends in a smaller bulb at the bottom. The bigger bulb is provided with a stopper having a lateral hole through which atmospheric connection can be made by turning the stopper, (*ii*) a graduated manometer fitted with the bulb, and (*iii*) a levelling tube connected

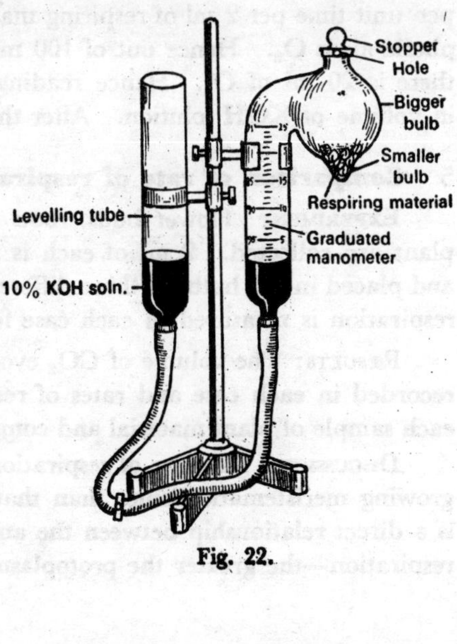

Fig. 22.

with the manometer tube by rubber tubing. The whole apparatus is clamped on a stand.

Two ml of respiring material (measured by displacement of water) like germinating seeds or flower petals are placed into the bigger bulb of the respirometer. A 10% solution of KOH is taken in the manometer tube. At the beginning of the experiment the air around the material is brought to the atmospheric pressure by turning the stopper of the bulb until its hole coincides with that of the neck of the bulb. The levelling of the reservoir tube on the right is so adjusted that the KOH solution in the tube is at the 100 ml mark at the bottom of the manometer. Two ml of respiring material is now surrounded by 100 ml of air. The experiment is started by turning the glass stopper at the top and thus cutting off connection with the atmospheric air.

RESULTS: As the respiration takes place in a closed space, the solution in the manometer tube rises up gradually. The reading should be taken up to 80 ml mark, i.e., up to 20 ml volume (since atmospheric oxygen is 20%) at an interval of 10 minutes, each time bringing the liquid in both the tubes at the same level, i.e., the liquid in the closed tube is brought under atmospheric pressure. Results are expressed as millilitre of CO_2 evolved per minute by the given respiring material.

DISCUSSION: The released CO_2, on coming in contact with KOH solution, is absorbed by it, oxygen is consumed and as a result KOH solution rises up in the manometer tube. The rate of rise of KOH solution is taken as a measure of rate of aerobic respiration in terms of volume of O_2 consumed per unit time per 2 ml of respiring material. One-fifth volume of atmospheric air is O_2. Hence out of 100 ml of enclosed air within respirometer there is 20 ml of O_2. Hence reading should be taken up to 20 ml rise in volume of KOH solution. After that anaerobic respiration will start.

5. Comparison of rate of respiration in different plant parts

EXPERIMENT: Flower buds, roots and leaves of a suitable herbaceous plant are collected. 2 ml of each is measured by displacement of water and placed in the bulbs of three different Ganong's respirometers. Rate of respiration is measured in each case following Expt. 4.

RESULTS: The volume of CO_2 evolved at an interval of 10 minutes is recorded in each case and rates of respiration are graphically plotted for each sample of plant material and compared.

DISCUSSION: The rate of respiration is always higher in young actively growing meristematic tissues than that of older and mature parts. There is a direct relationship between the amount of protoplasm and the rate of respiration—the greater the protoplasm, the higher is the respiration rate.

The hydration of protoplasm and quantity of respiratory enzymes are always greater in young cells compared to mature and vacuolated cells. Hence the respiratory rate is always higher in young cells which are rich in protoplasm. Vigorous respiration of root takes place if the space surrounding the roots and root hairs has ample supply of oxygen. In this experiment the maximum rate of respiration is expected in case of flower buds than in roots and leaves.

N.B. This experiment may also be performed with different types of seeds (starchy, proteinaceous and fatty).

6. Quantitative estimation of CO_2 evolved during respiration

(*a*) BY BARCROFT-WARBURG'S CONSTANT VOLUME MICRO-RESPIROMETER

PRINCIPLE AND DESCRIPTION: A convenient method of measuring the respiration of tissues in minute quantities has been developed by Warburg (1926). The tissue whose respiration is to be measured is placed in a closed container with an attached manometer which records changes in gas pressure as a result of oxygen consumption or carbon dioxide production.

The apparatus is shown in Figure 23. Each respirometer consists of two main parts, a glass flask *f* and a manometer *m*, separable at a ground glass joint *j*. The tissue is placed in the flask *f*. When only oxygen consumption is to be measured, $Ba(OH)_2$ or NaOH solution is added to the well *w* at the centre of the flask. But when both oxygen consumed and carbon dioxide produced are to be measured, an HCl solution is placed in the side arm *a* with stoppered neck *n* in addition to the alkali in *w*. The manometer fluid is contained in a rubber bulb *b* and can be added to or withdrawn from the manometer by adjusting the screws. This enables one to return the right side of the manometer to the starting point during making a reading and also to read the pressure on the left arm of the apparatus at constant volume. A mirror behind the manometer reduces parallax in reading. The manometer is provided with a two-way tap *t* at the closed end. The reaction chamber is kept in a bath of constant temperature. The entire apparatus is shaken to facilitate gas exchange and temperature equilibrium.

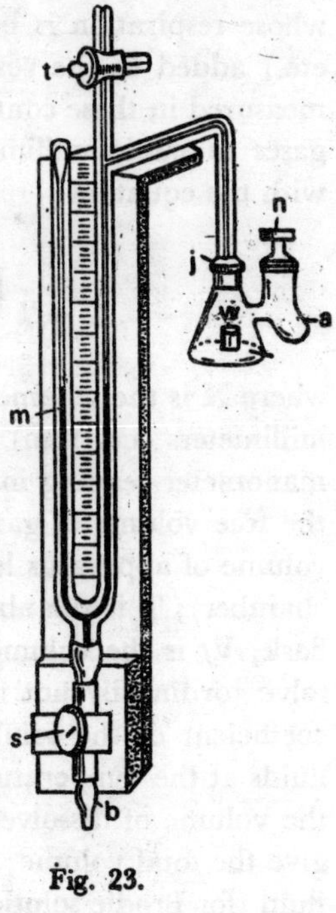

Fig. 23.

STANDARDIZATION: It is necessary to know the volume of the apparatus including the manometer to the manometer fluid in order to calculate gas volumes from changes in pressure. The manometer is detached and filled with clean mercury from the 15 cm mark to a marked point about 2 cm above the flask connection. Now mercury is poured into a beaker and the dry reaction chamber *f* is filled with clean mercury until it just rises to the marked point on the manometer when the manometer and flask are connected. This mercury is collected in a beaker and the temperature and weight of the metal are determined. The weight of the mercury in milligrams divided by its density at the observed temperature ives the volume of the apparatus in cubic millimeters.

Brodie solution (23 gm NaCl, 5 gm sodium cholate, 500 ml water, 5 drops conc. thymol in alcohol, as a preservative, and a few crystals of neutral red to colour) is used in the manometer to increase the sensitivity of readings, and to avoid sticking and other difficulties (see Appendix II). This solution has a density of 1·0336 and gives a manometric pressure of one atmosphere (760 mm Hg) at 10,000 mm. If in addition to these two values, the volume of the apparatus and the normal barometric height of the manometer fluid, we know the temperature, the volume of the material whose respiration is being measured, the volume of fluids (water, NaOH, etc.) added to the reaction chamber, and the solubility of the gas being measured in these contained liquids, the change in volume of the contained gases in cubic millimeters under standard conditions can be calculated with the equation

$$X = h\left(\frac{Vg\dfrac{273}{T} + Vf \times \alpha}{Po}\right)$$

where X is the volume of gas absorbed (− X) or evolved (+ X) in cubic millimeters (cu. mm) under standard temperature and pressure; *h* is the manometer reading in millimeters (reading of left arm minus right); V*g* is the free volume of gas in flask and manometer to manometer fluid (total volume of apparatus less volume of sample, liquids, etc., added to reaction chamber); T is the absolute temperature of the water around the reaction flask; V*f* is the volume of all fluids in which the measured gas might dissolve (ordinarily not including volume of solid samples); α is the Bunsen coefficient of the solubility of the gas being measured in the contained fluids at the temperature T (see below); it is to be noted that $Vf \times \alpha$ gives the volume of dissolved gas and that this is added to the free gas (V*g*) to give the total volume; P*o* is the normal pressure in terms of the manometer fluid (for Brodie solution 10,000 mm).

In its simplest terms the equation states that the change in gas volume during the experiment is equal to the fractional change in pressure *h*/P*o* times the total volume of the gas **V*g***, with corrections for temperature and the solubility of the gas in the fluids present.

THERMOBAROMETER: The above equation assumes that the barometric pressure, the temperature, and the vapour pressure of the contained liquids remain constant during the experiment and therefore cancel out. In practice, the great sensitivity of the manometer makes it necessary to set up an apparatus with the liquids, but without a sample, and to correct the experimental readings by the changes in the manometer of this blank apparatus, which are due to temperature or to barometric variations during the course of the experiment.

EXPERIMENT: This method is most suitable for measuring respiration of minute quantity of respiring material. The reservoir *b* of the manometer is filled with Brodie solution. 1 ml of the respiring material is placed in the outer part of the flask *f*. 0·2 to 0·4 ml of CO_2 free 2N KOH solution is taken in the central cup *w* and 0·3 to 0·6 ml 2·5 N HCl in the side arm *a*. The manometer connections are greased lightly but uniformly. The flask is secured in place with springs or rubber bands keeping the stopcock *t* open. The temperature of the water bath is kept constant at 30°C. If only O_2 consumption is to be measured, the HCl is omitted from the side arm and one or more samples are set up as desired in different flasks. If both O_2 and CO_2 are to be measured to obtain RQ, the HCl is included in the side arm and all experiments are set up in duplicate. In either case a flask is set up with a sample but with KOH and other fluid to serve as thermobarometer.

The stopcock *t* is left open. The assembled manometers and flasks are set in the water bath and shaken for 15 minutes to attain temperature equilibrium. Now with screw *s* the right arm of the manometers is adjusted to 250 point for which they are calibrated (150 to 250 mm), the stopcocks are closed and the time is recorded as the beginning of the experiment. When CO_2 production is to be measured, the duplicate flask for each sample is quickly removed, a finger is held tightly over the open end of the manometer to prevent the manometer fluid being blown out or sucked into the flask, the flask is tipped to mix the KOH and HCl solutions thoroughly from the cup *w* and arm *a*. These flasks are returned to the bath the manometer is carefully released, shaken for 5 to 8 minutes and manometer reading is recorded. This is corrected by changes in the thermobarometer, as CO_2 present or produced before experimental time.

The remaining flasks are shaken for 100 to 130 oscillations per minute for one hour or more. Intermediate readings for O_2 consumption may be made as desired by stopping the shaker and adjusting the right arms of the manometers to the original point and recording the manometer readings including that of the thermobarometer.

At the end of the experiment, the total O_2 consumption is recorded and absorbed CO_2 is liberated by the method used for the control samples taking care to protect the manometer fluid against changes in pressure and to bring the gases at water bath temperature with vigorous shaking before reading the CO_2 pressure. The change in pressure upon the mixing of KOH and HCl gives the '*h*' reading for CO_2. The right side of the manometer arm is always returned to its original setting before taking a reading since all of the calculations are based upon a constant volume of gas within the apparatus. The thermobarometer pressure is always recorded along with each reading.

CALCULATIONS: The proper terms are substituted in the equation

$$X = h\left(\frac{Vg\frac{273}{T} + Vf \times a}{Po}\right).$$

The gas absorbed or evolved is calculated in cu. mm, or in ml per gram of dry tissue per hour. The value V*g* varies with the flask and with the volume of added samples or other fluids. The volume of bacterial cultures and of KOH and HCl solution is obtained by pipetting; the volume of seeds, tissues, etc., by displacement. V*f* is usually the volume of sample and other fluids for bacterial cultures but does not include the volume of seed tissue.

In experiments in which a **constant volume** of sample is run at constant temperature, the value of the quantity within the brackets remains constant and can be assigned a value K (**flask constant**) so that the equation becomes X = *h*K, in which K is calculated for each flask at each temperature. The result may be expressed in mg by multiplying the volume of CO_2 in ml by the density of CO_2 at a particular temperature and pressure.

Table 3. Density of CO_2 and Bunsen coefficient (a) at different temperatures.

TABLE 3.

Density of CO_2

Temperature	Density at 760 mm
16°C	1·8267
20°C	1·7909
25°C	1·7443
30°C	1·6949

Value for Bunsen coefficient (a)

Temperature	CO_2	O_2
15°C	1·002	0·03456
20°C	0·901	0·03137
25°C	0·772	0·02874
30°C	0·648	0·02646

The effect of temperature or other factors upon the respiration and RQ of seeds and plant tissues may be measured by this apparatus.

(*b*) By Pettenkoffer's gas stream method of CO_2 estimation

Principle and description of apparatus: The principle of this method is that CO_2 liberated by respiratory tissue is removed from its chamber along with CO_2 free gas stream and absorbed in baryta (Barium hydroxide solution) taken in Pettenkoffer's tube to form barium carbonate. This is then titrated by a standard acid (HCl) to know its CO_2 content. CO_2 free air is drawn through a system as shown in Figure 24.

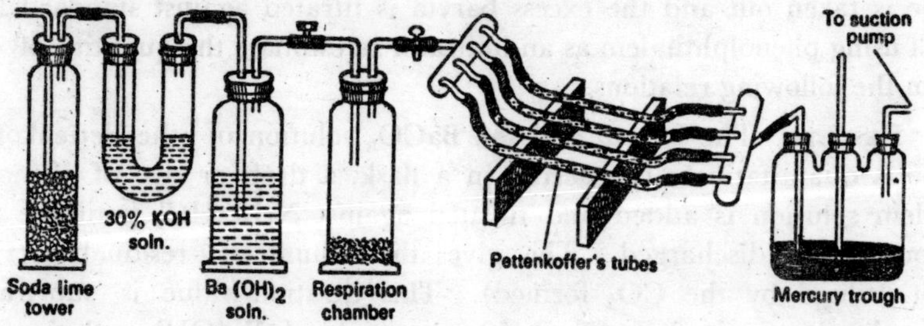

Fig. 24.

Description: At the extreme left there is a soda lime tower containing soda lime for absorbing CO_2 of the air entering through its opening at the mouth. The end of the tower is fitted with a tube through which CO_2 free air passes into a U-tube containing KOH solution (30%). The U-tube in turn is connected with another tower containing lime water or $Ba(OH)_2$ solution which is connected to the respiration chamber by means of a tube. The respiration chamber is again connected to one end of the horizontally placed Pettenkoffer's tube containing 50 ml of standard N/10 $Ba(OH)_2$ solution (8·567 gm/litre). The other end of this tube is connected to an aspirator or suction pump. Thus on applying a suction at the extreme right, air current enters the system of towers and tubes through the inlet at the top of the soda lime tower. A mercury trough may be introduced in between Pettenkoffer's tubes and the aspirator to regulate air flow. All connections should be made air-tight.

Experiment: At the beginning of experiment, weighed amount of plant tissue is taken in the respiration chamber. All the towers and Pettenkoffer's tubes are connected as described and made air-tight. The inlets of KOH tube and baryta tower must be dipped into the solution but the exit tubes should remain well above the surface of the solution. Now the aspirator or suction pump is started. Air is sucked out through the end of the Pettenkoffer's tube at the aspirator end causing the air to bubble into the solution of baryta contained in the Pettenkoffer's tube through the respiration chamber and other towers successively. Thus air first comes through the soda lime tower and then through KOH tube and baryta tower, thus becoming completely free of CO_2. This CO_2 free air containing O_2 comes in contact with the plant tissue in the respiration chamber and aerobic respiration takes place as a result of which CO_2 is evolved. This CO_2 produced by plant tissue during aerobic respiration then passes through the standard baryta solution of Pettenkoffer's tube and is completely absorbed by it forming $BaCO_3$.

After allowing respiration to occur for a particular time Pettenkoffer's tube is taken out and the excess baryta is titrated against standard N/10 HCl using phenolphthalein as an indicator to estimate the quantity of CO_2 from the following relations.

RESULTS: The baryta and the $BaCO_3$ solution of the Pettenkoffer's tube is quantitatively transferred in a flask, a drop or two of phenolphthalein solution is added and titrated against N/10 HCl until the pink colour is just discharged. This gives the volume of residual $Ba(HO)_2$ (not utilised by the CO_2 formed). This titration value is subrtracted from the titration value of fresh 50 ml sample of $Ba(OH)_2$ solution and weight of CO_2 liberated in respiration is calculated from the following equation:

$$mg\ CO_2 = V \times N \times 22$$

where V is the difference between blank and experimental titration values in millilitres, N is the normality of acid used in titration and 22 is the equivalent weight of CO_2 in $BaCO_3$. Since molecular weight of CO_2 is 44 and it is absorbed by $Ba(OH)_2$ as H_2CO_3 the equivalent weight is to be calculated by dividing the molecular weight by 2.

DISCUSSION: This method sometimes called the "gas-stream method", has the advantage of accuracy and convenience and of permitting a study of the cell material for an indefinite period under constant conditions. Here the CO_2 yield is obtained in milligrams, the result may be divided by thousand to give grams or divided by 1·977 (density of CO_2) to change to ml of CO_2 at O°C and 760 mm. Hg. It is to be noted that CO_2 is calculated directly rather than as the carbonic acid which is actually measured.

7. Determination of respiratory quotients (RQ) by Ganong's respirometer

EXPERIMENT: From this experiment, volume of CO_2 evolved and O_2 consumed during aerobic respiration can be directly obtained and RQ, i.e., (volume of CO_2/volume of O_2) can be calculated. The experiment can be conveniently performed by Ganong's respirometer described in Expt. 4. The manometer tube and reservoir are filled with brine solution (saturated NaCl solution) and by adjusting the reservoir, the surface of the brine solution is brought to 100 ml mark of manometer. The weighed equal quantities (2 ml) of tissues whose RQ is to be determined are taken in the bulbs of two similar types of Ganong's respirometer. In the bulb of one respirometer 1 ml of 40% KOH solution contained in a small vial is kept along with the tissue. Total volume of tissue plus KOH solution plus vial should be noted (let it be *k* ml). The other respirometer contains only tissue in the bulb and its volume is noted (let it be *t* ml). Now the air within the bulb is brought to atmospheric pressure by turning the stopper. The level of brine solution is brought to 100 ml mark at the bottom of manometer. This is done in both the respirometers and the connection with the outside air is cut off by turning the stoppers. Now the bulb and the manometer tube up to the level of brine solution contain a definite volume of air having 20% O_2 and 0·03% of CO_2 on an average. The set up is placed in dark or covered with black paper for a certain period of time (say two hours) and any change in the reading of brine level is noted. The volume of air in case of the respirometer containing plant tissue only is $100-t$ and this multiplied by 20/100 gives the volume of O_2 in this air. Similarly the volume of air in case of the respirometer containing KOH vial and the plant tissue is $100-k$ and this multiplied by 20/100 gives the O_2 in this closed atmosphere.

During the time for which the set up is kept, respiration occurs in the tissues of both the respirometers by absorbing O_2 and giving out CO_2. In the respirometer containing KOH vial, CO_2 given out is absorbed by KOH solution but this does not happen in the respirometer without KOH_1. Therefore, the brine level rises in the respirometer with KOH and either rises or falls or remains stationary in the respirometer without KOH. In the respirometer containing KOH, CO_2 is absorbed reducing the pressure in the manometer tube and brine level rises because O_2 has been used up by the tissue. The more O_2 is used up, the more rises the brine level. Now by adjusting the reservoir the brine level is brought to the same level in both the arms, thus bringing the volume of air in the manometer tube in atmospheric pressure. At this stage the brine level in manometer tube is noted and the difference between the final and initial readings gives the volume of O_2 used by the tissue (let it be *x* ml).

Now in a respirometer without KOH, O_2 is also used up giving out CO_2. But in this case CO_2 is not absorbed since KOH is not present. If CO_2 given out is less than O_2 used (when fat is a substrate) the brine level will rise in the manometer tube. If CO_2 given out is more than O_2 used (when acid is a substrate) the brine level will fall down in the manometer tube and when CO_2 given out is equal to O_2 used up (when carbohydrate is a substrate), the level of brine remains stationary. Before taking readings the brine level is adjusted at the same level in both the arms. The difference between the final and initial readings gives the net gas exchange which has taken place (let it be y ml). If the rise of brine level upward is considered as negative change and fall of brine level downward is considered as positive change, the volume of CO_2 evolved will be $(x-y)$ ml when brine level rises up and $(x + y)$ ml when brine level falls down. It may also remain stationary when RQ is unity.

$$RQ = \frac{\text{volume of } CO_2 \text{ evolved}}{\text{volume of } O_2 \text{ used}} = \frac{x+y}{x} > 1 \text{ or } \frac{x-y}{x} < 1 \text{ or } \frac{x}{x} = 1.$$

DISCUSSION: $\frac{x+y}{x}$ (> 1) will be the RQ when the respiring substrate is acid and $\frac{x-y}{x}$ (< 1) when the respiring substrate is fat. RQ will be unity when the brine level in the respirometer without KOH remains stationary indicating that the respiring substrate is carbohydrate. In this way the RQ for any given tissue may be calculated. It is to be particularly noted that the volume of the closed air within the bulb and the manometer should be equal in both the cases. The respirometers should be of the same size and volume and equal volumes of tissue should be taken in each case. Since KOH vial is kept in one bulb only, a similar vial containing equal quantity of brine solution may be kept in the other bulb so that same volume of air is present in both the bulbs. Temperature greatly affects the volume of gases and hence both respirometers should be placed at the same temperature.

8. Demonstration of liberation of heat energy during respiration

EXPERIMENT: Three thermosflasks are taken. One contains water-soaked seeds, second dry seeds and the third boiled seeds (dead) and all the lots are of equal weight. The mouth of the thermosflasks are corked through which passes a thermometer in each flask so that the bulb of the thermometer remains within the seeds. The flasks are left for 24 hours.

OBSERVATION: Rise of temperature is observed in all the cases.

INFERENCE: The rise in temperature of the flasks containing soaked seeds indicates that heat is produced during respiration of seeds. Temperature remains nearly unchanged in dry seeds and completely unchanged in case of boiled seeds. This is because the respiration of dry seeds is almost negligible and in case of boiled seeds nil.

N.B. The mathematical evaluation of heat loss during respiration of one gram mole of glucose is as follows. Every mole of ATP formed from ADP and phosphate requires about 12 K. Cal. of energy. But complete combustion or chemical oxidation of one mole of glucose yields 684 K. Cal. of energy as heat. Therefore, 684 − 456 = 228 K. Cal. is lost as heat energy. Thus in aerobic respiration of each mole of glucose about $\frac{456 \times 100}{684}$ = 67% energy of each glucose mole is carried into 38 mole of ATP. This is called as the "efficiency of energy conservation" in aerobic respiration.

9. Demonstration of loss of weight in respiration

EXPERIMENT: About 50 seeds are surface sterilised with 1% $HgCl_2$, washed thoroughly and 5 such seeds are separately placed in ten petridishes containing soaked filter paper. The pertidishes are kept in dark and seeds are allowed to germinate. The fresh and dry weights of each lot of seedlings are taken every alternate day.

RESULTS: The percentage loss in dry weight is calculated in each case and the results are graphically plotted taking loss in weight as ordinate and days as abscissa.

DISCUSSION: Respiration is a catabolic process which takes place by breaking down of stored carbohydrate and liberating CO_2, H_2O and energy. Since the seedlings are grown in dark the anabolic process, i.e., photosynthesis, cannot take place and as a result of which loss in weight occurs as the plants grow.

10. Effect of wounding on respiration

EXPERIMENT: Experiment can be conveniently performed either with Warburg's method or Pettenkoffer's method. Two lots of potato tubers are taken, one lot is slightly larger than the other. The larger potatoes are peeled off and surface area is made comparable with the other lot. Approximately the equal weights of these two lots are washed and dried and the two samples are placed in two respiration chambers of the Pettenkoffer's apparatus or in the flasks of Warburg's apparatus. The rate of respiration is determined in both the cases at a constant temperature.

RESULTS: Respiration rates in millilitres of CO_2 per gram dry tissue per hour are plotted. The dry weights of the samples are determined at the end of the experiment.

DISCUSSION: Injury to a given plant tissue often causes the respiratory activity to increase. Generally, in case of injured plant tissues the rate of respiration increases for the time being and this increase gradually rises to a maximum point and then the rate decreases. Wounding generally initiates meristematic activity in the area of the wound, resulting in the development of "wound callus". It has been shown that a considerable increase in sugar content (about 70%) around the injured cells takes place. Perhaps the increase in respiration due to wounding is caused by increased availability of respiratory substrate and meristematic cells.

11. Effect of pretreatment with carbon dioxide on the rate of respiration

EXPERIMENT: Equal weights of three lots of potato tubers are taken. Two lots are taken in two petridishes placed on two ground glass plates and each is covered with a 1-litre bell jar having an outlet at the top (Figure 25). The lower rims of the bell jars are made air-tight with grease. The bell jars are partially evacuated with the help of a suction pump. Now CO_2 is passed from a Woulfe's bottle ($CaCO_3+HCl$) into one bell jar for 10 minutes and to the other for 20 minutes. The outlets are then closed and tubers are kept in these atmospheres for about three hours. The control lot is also placed under a bell jar for the same period of time under normal atmosphere. A centrally placed thermometer records the temperature.

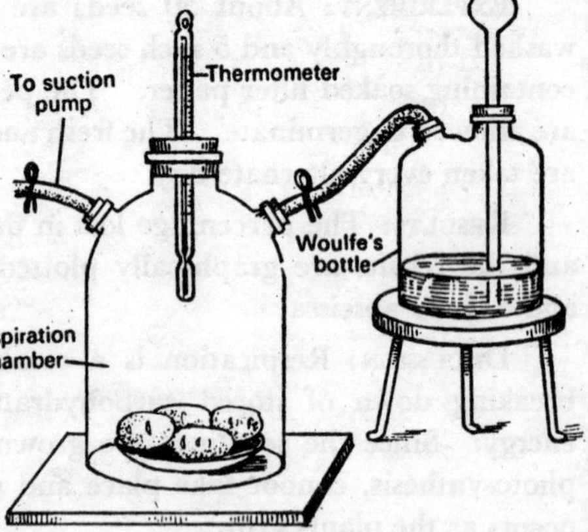

Fig. 25.

After the stipulated time the tubers are taken out and rates of respiration are determined with the help of Ganong's respirometer.

RESULTS: The rate of respiration in each case is graphically plotted and compared.

DISCUSSION: Increasing concentration of CO_2 has a definite repressing effect on respiration. Since the direct measurement of respiration rate in terms of CO_2 evolution and simultaneous increase in CO_2 concentration is difficult with ordinary apparatus, here the effect of pretreatment of CO_2 is studied. CO_2 input raises the internal concentration of CO_2 considerably and limits respiration by its toxic effect.

12. Effect of moisture content on respiration of grains

EXPERIMENT: About 300 grams of dehusked rice or wheat seeds are taken in a beaker and dried in an oven at 40°C for 24 hours. These seeds are divided into 5 lots. The initial moisture content of one lot is determined. Water is added to the other 4 lots and each lot is soaked for 5,

10, 15 and 30 minutes separately. The moisture content of each lot is then determined by taking a portion of seed from each lot. Equal weights of seeds from each lot including unsoaked one are taken in Ganong's respirometer and rate of respiration is measured.

RESULTS: The rate of respiration is expressed in terms of millilitre of CO_2 evolved per gram of dry seeds per hour and graphically plotted in each case.

DISCUSSION: Within certain limits, moisture content of tissue affects its respiratory rate. In dormant seeds water content is generally less than 10% and their respiratory rate is very slow. Seed. when come in contact with moisture imbibe water and swell and their respiratory rate gradually increases. The moisture content of the tissue increases the amount of soluble respiratory substrate and also the activity of the protoplasm by the enzymes.

13. **Effect of food supply on respiration.**

EXPERIMENT: One lot of leaf sample is collected from rice plants which were previously kept in dark for 24 hours and another lot is collected from rice plants previously kept in light for 24 hours. The rate of respiration of equal weights of each sample is measured in a Pettenkoffer's tube or in Ganong's respirometer. The dry weight of each sample is determined.

RESULTS: The results are expressed as millilitre of CO_2 evolved per gram of dry tissue per hour.

DISCUSSION: The rate of respiration of a given tissue is governed by the concentration of the soluble respirable substrates. Since organic materials are oxidised during respiration, the amount and kind of materials present in the cells appreciably affect both the rate and course of respiration. The rate of respiration increases due to increased carbohydrate production as a result of photosynthesis and decreases due to lower carbohydrate content in the dark. Thus the concentration of respirable substrate may limit the rates of CO_2 production or O_2 uptake.

N.B. Increased respiration is also observed when various sugars (especially sucrose, glucose, fructose or maltose) are supplied to floating leaves or other tissues in solutions.

14. **Determination of rates of respiration and nature of substrates by McDougal respiroscopes**

EXPERIMENT: This is a simple apparatus fixed in a wooden frame (Figure 26) which consists of a pair of vertical tubes. The upper end of the tube is wider than the lower end which is graduated. The lower end of the graduated tube dips into a beaker containing brine solution. The upper bulb-like wider ends of the respiroscopes are closed by means of corks through which pass two small tubes having bent ends. The upper end, i.e., the outside ends of the bent tubes are fitted with stopcocks. From the lower bent ends of the two tubes, two small vials one containing KOH pellets hang within the bulbs. Equal amounts of germinated starchy or fatty seeds (seed coat removed) are placed within the wider bulbs of the respiroscope on a soaked cotton plug. A thermometer may be centrally placed to record temperature.

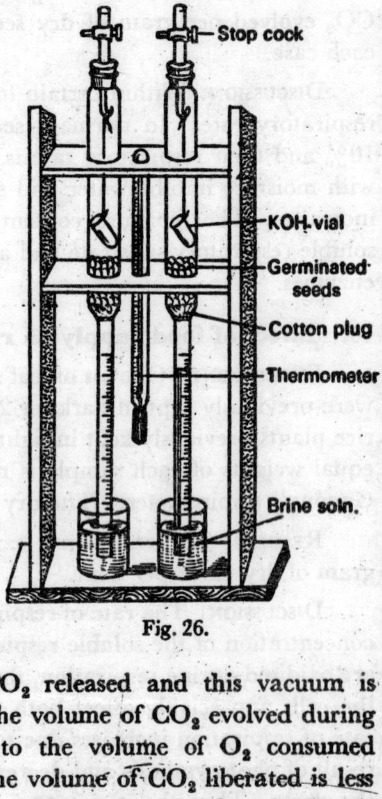

Fig. 26.

The stopcock is then turned to make connection with atmospheric air. The brine solution rises through the graduated end of the respiroscope and becomes stationary at the brine level of the beakers. The stopcock is closed and the level of brine solution in both the tubes is recorded. The rise of brine columns in the tubes is recorded at an interval of 15 minutes for 2 hours.

RESULTS: The rate of rise of water column indicates the rate of respiration.

DISCUSSION: The partial vacuum created due to absorption of O_2 during respiration cannot be filled up by CO_2 released and this vacuum is then filled up by the brine solution. The volume of CO_2 evolved during respiration by starchy seeds is equal to the volume of O_2 consumed (RQ = 1). But in case of fatty seeds the volume of CO_2 liberated is less than the volume of O_2 used up (RQ < 1). Hence the rate of respiration as measured in terms of O_2 consumption is less in case of fatty seeds than starchy seeds. The depression (when RQ > 1) or elevation (when RQ < 1) in the column of brine solution in the control McDougal respiroscope (without KOH) indicates the nature of substrate. If the brine level is stationary (RQ = 1) then the substrate is carbohydrate.

15. Respiration of roots of intact plants

EXPERIMENT: The adventitious roots of rice or wheat plants are carefully washed with water and placed in a bottle containing water which is made slightly alkaline with dilute NaOH solution. This is coloured pink with addition of a few drops of phenolphthalein. A second bottle is prepared in the same way, stoppered tightly and left without a plant. Both the bottles are allowed to stand in diffused light and the solutions are examined from time to time and change in colour is carefully noted in each case.

OBSERVATION: The pink colour of the solution in the bottle containing the plant gradually fades and ultimately becomes colourless after a considerable time. The solution of the other bottle remains as such.

INFERENCE: During respiration of roots CO_2 is released which is converted to H_2CO_3 when comes in contact with water. This acid neutralises the dilute NaOH solution and pink colour of the solution fades. When a portion of this neutralised solution is gently boiled for few minutes the pink colour reappears because CO_2 comes off on boiling leaving the solution alkaline again.

16. Demonstration of continuity of intercellular spaces

EXPERIMENT: A conical flask is taken and half-filled with water. The mouth of the flask is fitted with a cork having two holes, through one of which is inserted a long petioled leaf of arum (*Colocasia*) so that the cut end of the petiole remains well under water. Through another hole is inserted a bent glass tube which is connected to a suction pump. The end of this bent tube remains well above the water surface. All connections are made air-tight. The air within the flask is drawn out by the suction pump.

OBSERVATION: Bubbles are seen to come out from the cut end of the petiole into the water of the flask.

INFERENCE: The experiment shows that as the air is sucked, the atmospheric air enters through the stomata of the leaf and through the intercellular spaces ultimately comes out through the cut end of the petioles. This shows that stomata and intercellular spaces form a continuity and are involved in gaseous exchange.

N.B. A similar experiment can also be set up by introducing a woody leafless twig instead of arum. All the cut ends of the woody twig are sealed with paraffin wax excepting the lower end which dips in water. The atmospheric air in this case passes through the lenticels to the intercellular spaces and ultimately comes out in water through the lower cut end of the twig and submerged lenticels.

CHAPTER XI

PHOTOSYNTHESIS

Photosynthesis is a complex process in which carbohydrate is produced from relatively simpler compounds like CO_2 and H_2O by illuminated green cells of plants. Oxygen is formed and evolved as a by-product. In this process light energy (photons) absorbed by the chlorophyll of green cells is bound as chemical energy in the organic end-products (carbohydrate) and is thereafter available for life processes of plants and of animals which consume them. It is a process of extreme importance to all living beings, for, with a few minor exceptions (chemosynthetic bacteria), all life is dependent upon it.

Laws of thermodynamics postulate that the energy liberated in the destruction of organic matter, i.e., oxidation (respiration), is dissipated in the vastness of atmosphere and is lost beyond recovery. The liberated energy for all practical purposes becomes unavailable for the reverse conversion into chemical energy. Thus the energy required for organic synthesis must come from some external source and the only external source which is continuously available on the earth's surface is sunlight. About 35 per cent of the available light energy can be transformed into the chemical energy of the sugar produced. Every year some 200 billion (one million × one million) tons of carbon go through the photosynthetic process. It has been estimated that plants take up 7×10^{11} tons of CO_2 to produce roughly 5×10^{11} tons of solid plant material. Approximately 90 per cent of the world's photosynthesis is carried out, not by the land plants, but by marine and fresh water algae.

The energy obtained from light drives four types of reactions in the chloroplast: (*i*) The photolysis of water (the Hill reaction), (*ii*) photosynthetic phosphorylations, (*iii*) the fixation of carbon dioxide, and (*iv*) the synthesis of carbohydrate. The photolysis of water is apparently carried on the lamellae (*thylakoids*) within the *grana* and also probably the photosynthetic phosphorylations. The subsequent steps of carbon dioxide fixation and starch synthesis are reactions of the soluble enzyme systems in the *stroma* of the chloroplasts.

The over all process of photosynthesis can be represented by the following equation:

$$6\ CO_2 + 12\ H_2O \xrightarrow[\text{light energy}]{\text{chlorophyll}} C_6H_{12}O_6 + 6\ O_2 + 6\ H_2O$$

$$(\Delta F = +118\ \text{K. Cal./mol of } CO_2)$$

This also illustrates that the source of oxygen is water and not carbon dioxide.

Photosynthesis like any other physical and chemical process is affected by the conditions of environment in which it occurs. 'The law of three cardinal points' by Sachs states that there is minimum, optimum and maximum for each factor in relation to photosynthesis. That all factors of a physiological process, particularly the smallest one, affects the total rate of the process, was first realised by Liebig (before Liebig, C.S. Sprengel made clear statement of the 'law of minimum', wrongly credited by many authors—including Liebig himself) who gave the 'law of minimum'. Blackman elaborated Liebig's statement and enunciated the 'principle of limiting factors', which reads as follows, "when a process is conditioned as to its rapidity by a number of separate factors, the rate of the process is limited by the pace of the 'slowest' factor"

Four environmental factors are of primary importance in influencing the rate of photosynthesis: (*i*) light, (*ii*) concentration of CO_2 of atmosphere (generally 0·03 per cent), (*iii*) temperature and (*iv*) supply of water. Other environmental factors which influence the rate of the process, though not markedly, include: (a) oxygen, (b) various chemicals, (c) wounding or injury, and (d) supply of mineral salts. A number of internal factors which also affect the process can be cited: (*i*) chlorophyll content of the assimilating surface, (*ii*) hydration of the protoplasm, (*iii*) anatomy of the leaf, (*iv*) photosynthetic enzymes, coenzymes and ATP, and (*v*) accumulation of end products of photosynthesis within the cells.

1. Simple demonstration of photosynthesis

(*a*) WITH THE HELP OF A BEAKER AND A FUNNEL

EXPERIMENT: A large beaker of capacity 500 ml is taken and is filled two-thirds with distilled water containing 0·1% $KHCO_3$ which acts as a source of CO_2. Some fresh and healthy aquatic plants like *Hydrilla* are taken in a beaker and the plants are cut obliquely at their bases under water. Cut ends are tied together with the help of a thread and are kept towards the neck of an inverted funnel in such a way that the limb of the funnel almost covers the *Hydrilla* plants and the stem of the funnel remains about one centimeter under the water surface. The whole set-up is now exposed to bright light and observed from time to time. Another set-up is similarly prepared and kept under a very low light intenstiy.

OBSERVATION: It is observed that evolution of bubbles from cut ends of the plants takes place in the set-up exposed to light. Little evolution of bubbles takes place in the set-up maintained in low light intensity.

INFERENCE: In light, evolution of oxygen bubbles takes place due to photosynthesis. This is further proved by the fact that little evolution of bubbles takes place in the set-up placed in low light intensity.

(*b*) WITH THE HELP OF WILMOTT'S BUBBLER

EXPERIMENT: The apparatus consists of a flask of capacity 500 ml fitted with a rubber cork having a central hole through which passes a glass tube. The lower end of the tube reaches the middle of the flask while its upper end forms a jet within a cylindrical cup. A graduated tube having a stopcock at one end remains inverted over the jet (Figure 27).

Now cut ends of one or two *Hydrilla* plants are introduced within the lower end of the glass tube. The flask is filled with distilled water and a small amount of bicarbonate is added to it. The *Hydrilla* plants with the tube are immersed within the flask and the rubber cork is fitted tightly. The upper cylindrical cup is filled with water so that the jet remains well below the water level. The graduated tube filled with water is inverted over the mouth of the jet and clamped properly. Two such set-ups are arranged and one is kept in bright light and the other in low light.

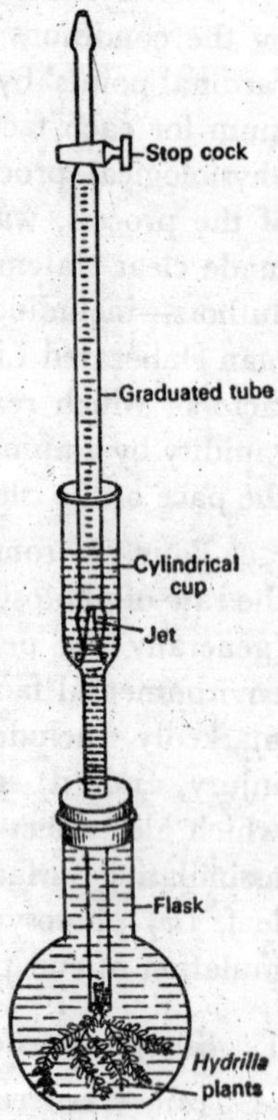

Fig. 27.

OBSERVATION: Same as in Expt. 1(*a*).

INFERENCE: Same as in Expt. 1(*a*).

N.B. The size of the bubbles can be adjusted conveniently by varying the oblique surface of the cut ends of the *Hydrilla* plants.

2. To study the "primary photochemical reaction" of photosynthesis

(*a*) TO STUDY PHOTOLYSIS OF WATER FROM EVOLUTION OF OXYGEN

(*i*) PHYSICAL PROOF OF OXYGEN EVOLUTION

EXPERIMENT: Two experimental set-ups are arranged as in Expt. 1(*a*). Now a test tube filled with water is inverted over the upper end of the stem of the funnel in each case, using the thumb to close the mouth of the tube while inverting.

Both the set-ups are placed in bright sunlight. As. photosynthesis occurs oxygen bubbles are given out, which rise up in water through the funnel stem into the inverted test tube and collect at its top by displacement of water. When the gas has collected to a considerable amount it is tested for oxygen. One of the gas-filled test tubes (other is kept aside for the next experiment) is removed by using thumb to cover its mouth and the thumb is then removed over a burning splinter or match stick.

OBSERVATION: The burning splinter or the match stick glows vigorously.

INFERENCE: Since the burning splinter or match stick glows in the collected gas, it evidently proves that the collected gas is oxygen.

(*ii*) CHEMICAL PROOF OF OXYGEN EVOLUTION BY ALKALINE PYROGALLATE (QUALITATIVE):

EXPERIMENT: The gas-filled test tube from Expt. 2(*a*) is taken and placed in a petridish containing potassium pyrogallate solution.

OBSERVATION: It is observed that the level of water gradually rises up in the test tube.

INFERENCE: Alkaline pyrogallate is a strong absorbent of oxygen gas. This absorbs the oxygen in the test tube which creates a vacuum, consequently water rises upward. Hence it proves that the evolved gas collected in the test tube during photosynthesis is oxygen.

N.B. If the test tube is half-filled with gas, water in the test tube must be carefully excluded from the tube and the mouth of the test tube closed with the thumb is then placed on alkaline pyrogallate solution.

(*iii*) CHEMICAL PROOF OF OXYGEN EVOLUTION BY INDIGOCARMINE SOLUTION (QUALITATIVE)

EXPERIMENT: A very dilute solution of the dye indigocarmine is prepared with previously boiled and cooled distilled water. Now minute quantity of sodium or potassium hydrosulphite ($Na_2S_2O_4$, $2H_2O$ or $K_2S_2O_4$, $2H_2O$) is very carefully added to the solution until the blue colour just disappears. (A dilute sodium hydrosulphite solution may be added dropwise from a burette and addition of the solution is stopped as soon as the colour just disappears. This will avoid the possibility of excess addition of the reductant). Two test tubes are completely filled with this reduced decolourised dye solution. A few fresh and healthy *Hydrilla* plants are put in each and corked carefully to exclude all contact with atmospheric air. One tube is exposed to bright light while the other is kept in dark.

OBSERVATION: The solution of the first test tube which was kept in light soon turns blue whereas the second one which was kept in dark remains colourless.

INFERENCE: Indigocarmine is auto-oxidisable, i.e., whenever reduced indigocarmine comes in contact with air or oxygen it becomes oxidised.

Indigocarmine which is reduced to a colourless compound by sodium hydrosulphite is oxidised back with molecular oxygen to its original state having blue colour in case of the first test tube where evolution of O_2 takes place due to photosynthesis. In case of the second test tube where to photosynthesis takes place, it remains as such.

(*iv*) CHEMICAL PROOF OF OXYGEN EVOLUTION BY WINKLER'S TEST (QUANTITATIVE)

EXPERIMENT: Three burettes are taken. One is filled with 40% manganous chloride solution, second one is filled with KI+KOH (dissolve 175 gm KOH and 37·5 gm KI in 250 ml H_2O) solution and the third one with conc. HCl.

Two Erlenmeyer's flasks (250 ml) are filled with 150 ml previously boiled and cooled (to free from oxygen) water. To each flask 5 ml of 1·35% $KHCO_3$ is taken. A few fresh *Hydrilla* plants or green alga like *Spirogyra, Chlorella* or *Scenedesmus* are put to one flask only and both are kept in bright light. After one hour 5 ml of water from each flask is taken separately in two 100 ml flasks. To each flask the following reagents are added one by one.

A. 0·5 ml of 40% manganous chloride.

B. 1·0 ml of (KI+KOH) solution.

(It is stoppered quickly and carefully mixed by rotating the flask.)

C. 2 ml of conc. HCl.

(The precipitate of manganous hydroxide is redissolved.)

The dissolved oxygen now liberates free iodine from the KI present. It is then titrated against 0·01 N sodium thiosulphate in both the cases using 0·5 ml of 0·5% starch solution till the blue colour disappears. Oxygen concentration in both the cases is calculated. During titration starch indicator should be used when a straw colour of the solution appears on titration with $Na_2S_2O_3$.

RESULTS: The oxygen concentration is calculated from the following relation. 1 ml of 0·01N sodium thiosulphate ($Na_2S_2O_3$, $5H_2O$) = 0·08 mg of oxygen or 0·0558 ml of oxygen. The initial oxygen concentration of water is calculated in both the flasks. The final concentration is again calculated in each case. Difference between the initial and final values gives the increase in oxygen concentration in the medium. The result may be expressed as milligram of oxygen evolved per gram weight of the plant (fresh or dry) or as millilitre of oxygen evolved per hour by constant volume of the plant.

DISCUSSION: During this experiment a number of chemical reactions takes place ultimately liberating iodine which is stoichiometrically equivalent to oxygen. In the last steps, the dissolved oxygen liberates chlorine which again liberates free iodine from KI present in the reaction mixture. The liberated iodine is then titrated against a standard sodium thiosulphate solution. This stands as an index of the oxygen content of the medium.

(*i*) $Mn(OH)_2+4HCl+O \rightarrow MnCl_2+3H_2O+Cl_2$

(*ii*) $2KI+Cl_2 \rightarrow 2KCl+I_2$

(*iii*) $I_2+2Na_2S_2O_3 \rightarrow Na_2S_4O_6+2NaI$.

N.B. If the concentration of $Na_2S_2O_3$ used be 0·025(N) and a 200 ml aliquot of the sample is titrated, then the relation is: millilitre of $Na_2S_2O_3$ $\equiv$ ppm dissolved oxygen.

(*b*) TO STUDY PHOTOLYSIS OF WATER BY DEMONSTRATION OF HILL REACTION

EXPERIMENT: About 25 gm of spinach or lettuce leaves are homogenised in 50 ml of ice-cold 0·5 M sucrose solution. It is then filtered through double-layered muslin cloth or glass wool. The suspension is centrifuged at low speed (500 rpm) for 10 minutes. The supernatant is recentrifuged at 4000 rpm for 10 minutes. The sediment is taken and resuspended in 0·5 M ice-cold sucrose solution. It is then kept in dark.

The following solutions are taken in three test tubes.

Test tube No 1 : 9 ml of phosphate buffer (pH 6·5) + 1 ml of chloroplast suspension (killed by keeping the test tube in boiling water) + 1 ml of 0·1% 2, 6-dichlorophenol-indophenol dye.

Test tube No. 2: 9 ml of buffer + 1 ml of chloroplast suspension + 1 ml dye.

Test tube No. 3: Same as in No. 2.

All the test tubes are shaken and percentage of absorption is read in a colorimeter at 430 nm.

Tube 1 and tube 2 are then placed in light and tube 3 in dark. The percentage of absorption in each test tube is again measured at 15 minutes interval for 60 minutes. Before taking each reading, the test tubes should be shaken well.

RESULTS: The change in percentage absorption is noted in each case. The percentage of absorption decreases in case of tube number 2 due to reduction of dichlorophenol (reduced dichlorophenol is colourless). Whereas no change occurs in case of tubes 1 and 3.

DISCUSSION: The production of oxygen without accompanying CO_2 fixation by isolated chloroplast, illuminated in the presence of substances like dichlorophenol-indophenol, ferric oxalate, ferricyanide, etc., the so-called Hill oxidants, is known as the *Hill reaction*. The light energy absorbed by chlorophyll and accessory pigments is utilised in splitting up of water into hydrogen and oxygen. The oxygen is liberated as gas while hydrogen is absorbed by Hill oxidant which gets reduced. This is photolysis of water (photolysis differs from electrolysis in that hydrogen gas does not appear simultaneously with oxygen gas but is used to reduce some compounds present in the chloroplast). The present experiment shows that Hill reaction takes place in test tube number 2 where all the requirements of this reaction are fulfilled whereas in case of tube numbers 1 and 3, Hill reaction does not occur due to killing of chloroplast and absence of light respectively.

N.B. To prepare 6·5 pH buffer, 0·1 M disodium hydrogen phosphate and 0·1 M monosodium hydrogen phosphate are mixed in the proportion of 4 : 6.

$$\text{M. W. of } Na_2HPO_4, 2H_2O = 178{\cdot}16$$

$$\text{M. W. of } NaH_2PO_4, 2H_2O = 156{\cdot}00$$

3. To study the "dark reaction" of photosynthesis

(*a*) EVIDENCE FROM STUDY OF TEMPERATURE COEFFICIENTS (Q_{10}) AT HIGH AND LOW LIGHT INTENSITIES

EXPERIMENT: An experiment is set up following the procedure of Expt. 1(*a*) or 1(*b*). If Expt. 1(*a*) is followed, a glass jet having a slight bent at the tip is to be fitted to the short stem of the funnel by rubber tubing (proper care must be taken in drawing out the jet end from a glass tube, for, if the bending is not correct or if the jet opening is too small or too big, the evolved gas may accumulate there and does not come out at all. For convenience a dropper may be used as a jet).

A pinch of bicarbonate is added to the water of the beaker and the set-up is kept in bright sunlight. The temperature of water (say, X°C.) is noted by a thermometer. Now the number of oxygen bubbles evolved and coming out through the fine jet-end is counted per unit time preferably a minute (it is advisable to discard very small bubbles coming out in torrents, as it is almost impossible to count them and better to select only relatively bigger bubbles of more or less uniform size). Five such readings are taken and the average number of bubbles per minute at that particular temperature and light intensity is recorded. The temperature of water of the beaker is lowered to 10°C. below that of first temperature, i.e., (X−10)°C. using ice blocks. The bubbles are similarly counted at that temperature and light intensity and average number of bubbles per minute is recorded. The temperature of water is now raised to 10°C. above that of the first temperature, i.e., (X+10)°C. using hot water or hot water bath. The bubbles are counted as usual at that temperature and light intensity and the average number per minute is recorded.

The same experiment is repeated at low light intensity, i.e., not under direct sunlight (if artificial light is used, the source may be removed to a distance to lower the light intensity). Bubbles are counted at each temperature at low light intensity and average number of bubbles per minute is recorded separately.

RESULTS: If the number of bubbles per minute at light intensity and at temperature of (X+10)°C. is *a*, the number of bubbles at X°C. is *b*, and that at (X−10)°C. is *c*, then Q_{10} can be calculated as a/b and b/c. Average of these two gives the temperature coefficient of photosynthesis at that light intensity. Q_{10} is similarly calculated at low light intensity.

DISCUSSION: The temperature coefficient or Q_{10} of a chemical reaction or of a physiological process is the ratio of the rate of the reaction or process at a given temperature to the rate at a temperature 10°C. below it. For purely photochemical reaction Q_{10} is about unity and for ordinary chemical reaction Q_{10} is at least two.

When photosynthesis is proceeding rapidly in well illuminated leaves receiving an ample supply of CO_2, Q_{10} is always greater than two. When the light intensity is low Q_{10} value is much less than two and sometimes approaching unity, i.e., values characteristics of photochemical reactions are obtained. Thus in addition to at least one photochemical stage in photosynthesis there must be at least one chemical stage (dark reaction). When the light intensity is high the temperature-dependent dark stage is limiting; when the light intensity is low, the photochemical stage is limiting and the whole process becomes relatively independent of temperature.

N.B. As majority of biological activities are catalysed by enzymes the Q_{10} law (Van't Hoff's law) is applicable to these bioprocesses within the temperature range of 0°C. to 30°C. Above 30°C. reaction rate declines characteristically and the Q_{10} values decrease from two to nearly one. Thus Van't Hoff's law is not applicable at higher temperature, i.e., above 30°C., as enzyme inactivation takes place more rapidly at higher temperature. At higher temperature, i.e., above 30°C., the Q_{10} values must be carefully interpreted with particular attention to the effect of time. Because the rate of reaction may be doubled between 20°C. to 30°C. for a period of two hours, but if continued for several hours, the rate will decrease.

(*b*) EVIDENCE FROM EXPERIMENTS IN INTERMITTENT LIGHT

EXPERIMENT: Two experiments are set up as in Expt. 2(*b*). Initially both the set-ups are illuminated at the same light intensity and the rate of evolution of bubbles per minute is recorded. One is then kept in dark for 15 minutes with a bell jar covered with black paper. After 15 minutes the bell jar is taken off and the rate of evolution of bubbles in both the sets are measured for 5 minutes. The same set-up which was in dark is again kept in dark for another 15 minutes. The bell jar is again taken off and rates are measured in each set. The experiment is repeated and four or five such readings are taken.

RESULTS: The average rate in continuous light and in intermittent light is calculated and results are represented in bar graphs.

DISCUSSION: The amount of photosynthesis per unit time (continuous rate) shown by green cells exposed to continuous illumination is less than the amount of photosynthesis per unit time of illumination (intermittent rate) by green cells exposed to alternating periods of light and dark. The rate in intermittent light is increased as the light periods are shortened and some times become nearly double the rate in continuous light. Evidently a given quantity of light energy is used in photosynthesis to a greater extent when supplied, not continuously, but in succession of small amount during short light periods. This happens because an essential phase of photosynthesis (dark reaction) goes on independently of the presence of light. In intermittent light, this phase then proceeds in the dark periods as well as in light periods and consequently, the rate of photosynthesis per unit time of illumination is enhanced. Moreover, in continuous light the photosynthetic product may inhibit the rate whereas in intermittent light the product may be used up in the dark phase.

N.B. In case of terrestrial plants the rate of photosynthesis is measured by weight-area method (Ref. Expt. 4(*iii*) of Chapter XII).

4. To study the essentiality of the factors for the photosynthetic process

(*a*) LIGHT DEPENDENCE OF PHOTOSYNTHESIS

(*i*) BY NOTING EVOLUTION OF OXYGEN IN LIGHT AND DARK

EXPERIMENT: Experiment is set up as in Expt. 1(*a*) or 1(*b*). It is kept in light for some time and then placed in dark.

OBSERVATION: Only in light evolution of oxygen bubbles takes place. In dark no evolution of oxygen bubble is observed.

INFERENCE: Hence light is essential for photosynthesis, since without light no photolysis of water takes place and oxygen does not evolve.

(*ii*) BY USING GANONG'S LIGHT SCREEN

EXPERIMENT: Leaves of a suitable potted plant are made starch-free by keeping the plant in dark for 24 to 48 hours. A starch-free leaf from this plant is then covered with Ganong's light screen (Figure 28). The main advantage of this light screen is that a part of the leaf is cut off from the light without hampering its ventilation. The whole plant with the light screen attached to one of its leaves is then kept in bright light. After 24 hours the screen is taken off and the leaf is severed from the plant. It is then bleached with boiling water and then with 95% boiling alcohol. A few drops of lactic acid may be added for better bleaching. After its complete decolourisation, the yellowed leaf is washed with 1% iodine solution for about a minute or so.

OBSERVATION: The exposed part turns blue whereas the covered portion (where light could not penetrate) remains yellowish red.

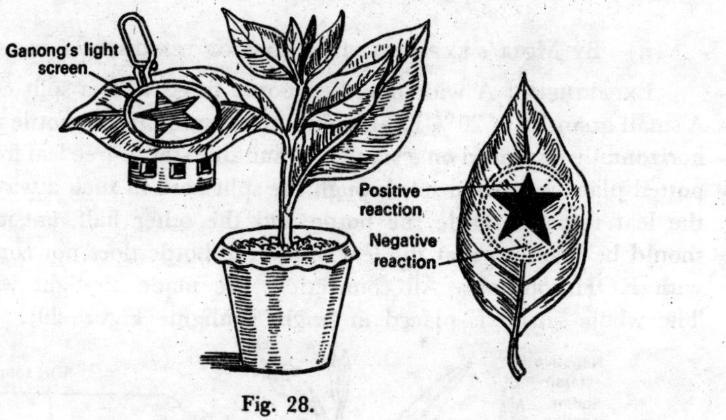

Fig. 28.

INFERENCE: The blue colour indicates the formation of starch—a product of photosynthesis. Yellowish portion shows absence of starch. Hence the experiment proves the necessity of light in photosynthesis.

N.B. Instead of a light screen a suitable photographic negative, which must be bright with clear-cut differentiation into black and white areas, can be used. After fixing the negative with sufficient care on a suitable starch free leaf and exposing the leaf with the negative in strong light for a day or two and subsequently treating the decolourised leaf with iodine solution, reasonably good positive print could be obtained.

(*b*) CO_2 DEPENDENCE OF PHOTOSYNTHESIS

(*i*) BY USING CO_2- FREE WATER (IN CASE OF AQUATIC PLANTS)

EXPERIMENT: An experiment is set-up as in Expt. 1(*a*) using previously boiled and cooled distilled water (in order to free it from dissolved CO_2). It is then kept in bright light and observation is made regarding evolution of O_2 bubbles. Now a pinch of $KHCO_3$ is added to the distilled water and allowed to dissolve in it completely. Observation is again made regarding evolution of O_2 bubbles.

Observation: No oxygen bubble comes out in CO_2-free distilled water but a number of oxygen bubbles comes out when a pinch of $KHCO_3$ is added.

Inference: The aquatic plants with their inherent adaptibility can effectively utilise dissolved CO_2 in water for photosynthesis because they have no contact with atmospheric air. As the boiled distilled water is devoid of dissolved CO_2, the *Hydrilla* plants cannot photosynthesise. But when this water is enriched with dissolved CO_2 by adding bicarbonate, these plants can easily photosynthesise and as a result O_2 bubbles come out.

(*ii*) By Moll's experiment (in case of terrestrial plants)

Experiment: A wide-mouthed bottle fitted with a split cork is taken. A small quantity of 20% KOH solution is poured in the bottle and clamped horizontally or placed on a stand. A suitable starch-free leaf from a healthy potted plant is introduced through the split cork in such a way that half of the leaf remains inside the bottle and the other half outside it. Care should be taken so that the leaf inside the bottle does not come in contact with KOH solution. All connections are made air-tight with vaseline. The whole set-up is placed in bright sunlight (Figure 29). After 2 to 3

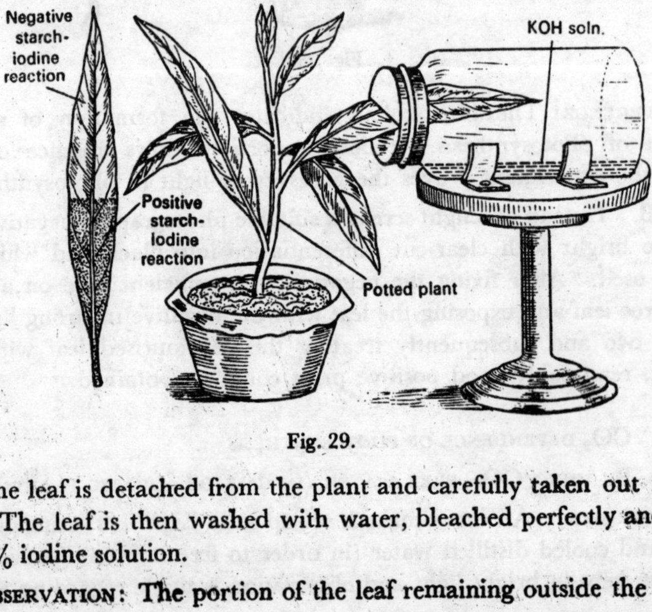

Fig. 29.

hours the leaf is detached from the plant and carefully taken out from the bottle. The leaf is then washed with water, bleached perfectly and treated with 1% iodine solution.

Observation: The portion of the leaf remaining outside the bottle coloured blue with iodine while the portion remaining inside the bottle takes no blue colour with iodine.

INFERENCE: The portion of the leaf remaining outside the bottle receives all conditions for photosynthesis (viz., CO_2, light and chlorophyll) and so starch formation takes place normally which is evident from the blue colour with iodine. The other half of the leaf which remains inside the bottle and where no CO_2 is available due to absorption by KOH solution, remains yellow showing no starch formation. This proves that photosynthesis cannot take place in an atmosphere completely devoid of CO_2.

(c) CHLOROPHYLL DEPENDENCE OF PHOTOSYNTHESIS

EXPERIMENT: A plant having variegated leaves (croton, *Coleus*, etc.) is selected. Variegated leaves have green chlorophyllous areas and nongreen areas dispersed on the leaf. Such a plant is allowed to starve in dark for 24 to 48 hours to deplete them of starch. An outline of a leaf from this plant is drawn on a paper and areas which are green and nongreen are traced on it. Now the plant is exposed to light for 24 to 48 hours. After this the variegated leaf is tested for starch by iodine.

OBSERVATION: It is seen that the presence of starch is shown only in green areas and not in nongreen areas.

INFERENCE: This experiment shows that photosynthesis can occur in leaf in its green areas only containing chlorophyll and without chlorophyll no photosynthesis takes place.

5. To study environmental factors controlling photosynthesis

(a) LIGHT FACTOR

(i) TO STUDY THE EFFECT OF LIGHT INTENSITY

EXPERIMENT: An experiment is set up as in Expt. 1(*a*) or 1(*b*). The set-up is placed at a distance of two meters from a strong artificial light (200-watt) without a reflector. The number of bubbles coming out per minute is counted. The distance between the set-up and the light source is reduced by one-eighth and the number of bubbles per minute is recorded. A few minutes should be allowed for adjustment to a particular light intensity by the plants. The distance is again reduced by one-fourth and one-half respectively of the original one and number of bubbles is recorded similarly in each case. The temperature should be carefully recorded every time.

RESULTS: The number of bubbles is plotted against the respective light intensity, keeping in mind that light intensity varies inversely as a square of the distance. The light intensity at the maximum distance may be measured, or it may be taken as unity, and the respective intensities of the shorter distances are calculated.

DISCUSSION: Generally the rate of photosynthesis is increased if the intensity of light is increased until some other factors, e.g., CO_2 conc., temperature, etc., become limiting. Within a limited range of intensity, the photosynthetic rate is more or less proportional to the light intensity, if the CO_2 concentration is not limiting. At high light intensity, the rate of photosynthetis is inhibited (solarization due to photo-oxidation). At a very low light intensity stomata are closed and hence entrance of CO_2 is checked and consequently the rate of photosynthesis decreases. Again as a result of very high light intensity transpiration rate increases causing reduction of water content of the cells of the leaf and thereby retarding photosynthetic rate. Moreover at high light intensity photo-oxidatian of chlorophyll takes place. Within a limited range, increase in light intensity also increases temperature which does not, therefore, become a limiting factor.

(*ii*) TO STUDY THE EFFECT OF LIGHT QUALITY

EXPERIMENT: The effect of light quality on the rate of photosynthesis can be studied by the same experimental set-up, keeping the light source at a constant distance and by separately using blue, green and red coloured cellophane papers as filter. The number of bubbles are counted in each case.

RESULTS: The results are plotted graphically taking quality of light as abscissa and number of bubbles as ordinate.

DISCUSSION: The effects of different light qualities are different on the rate of photosynthesis. The rate of photosynthesis is maximum at the red region of visible spectrum (655 nm) and secondary maximum in the blue region (440 nm). Thus two peaks are obtained, while a plateau is formed at the green region because most of light of this quality is reflected by the plant. The action spectrum thus shows that the maximum efficiency in photosynthesis is exhibited by the red and blue regions of the visible spectrum (400 to 700 nm).

(*b*) CARBONDIOXIDE FACTOR

EXPERIMENT: An experiment is set up as in Expt. 1(*a*) or 1(*b*) using 500 ml of previously boiled and cooled distilled water. Now six lots of 1·25 gm weight of $KHCO_3$ are taken separately. The set-up is placed in bright light. A graduated tube or a centrifuge tube filled with water is inverted over the jet. 1·25 gm of $KHCO_3$ is then added and allowed to dissolve completely. After waiting for 5 minutes, the volume of oxygen collected in the graduated tube is recorded. The remaining five weights of bicarbonate are separately added one by one allowing time to dissolve the bicarbonate completely and volume of O_2 collected after each addition of bicarbonate is recorded. After addition of each lot of $KHCO_3$ about 5 minutes should be allowed for complete dissolution and then the volume of O_2 is collected for a specific duration. The concentrations of the solution, after addition of first to sixth lot of $KHCO_3$ packets, will be 0·25, 0·50, 0·75, 1·0, 1·25 and 1·50 per cent respectively.

RESULTS: The results are graphically plotted taking concentration of bicarbonate as abscissa and volume of O_2 as ordinate.

DISCUSSION: An increase in the CO_2 concentration of the surrounding atmosphere results in an increase in the rate of photosynthesis until a point is reached where further increase in CO_2 concentration brings about no increase of photosynthetic rate. At this stage other factors like light or temperature become limiting. Again high CO_2 concentratian retards photosynthesis because of (*i*) increased acidity (HCO_3^- from bicarbonate and H^+ from water forming H_2CO_3) of the mesophyll cells, (*ii*) of a narcotic effect on the metabolic function of the cells, and (*iii*) closing of stomata.

N.B. In case of terrestrial plants effect of CO_2 concentration on the rate of photosynthesis (in terms of starch formation) may be studied by covering the plant material with a bell jar fitted with a CO_2-gas-measuring burette at the top to control the CO_2 inflow inside the belljar with respect to time.

(*c*) TEMPERATURE FACTOR

EXPERIMENT: The same procedure and set-up employed in Expt. 3(*a*) can be used for this experiment. In this experiment the temperature of water is raised by 5°C. instead of 10°C. Number of bubbles per unit time is recorded at each temperature. Readings should be taken in at least five different sets of temperatures (range of temperature may be kept between 10°C. to 50°C.).

RESULTS: Results are graphically plotted taking temperature as abscissa and rate of photosynthesis as ordinate.

DISCUSSION: Photosynthesis is restricted to a temperature range (0°C. to 60°C.) that roughly corresponds to that tolerated by protein compounds. Although the photochemical part of photosynthesis is independent of temperature, biochemical part, which is controlled by enzyme activity, is strictly temperature dependent. However, there appears to be a wide variation in adaptability amongst plants in their ability to tolerate temperature extremes. Generally above 35°C. temperature has a deteriorating effect.

N.B. Q_{10} may be calculated by knowing the rate of photosynthesis at every 10°C. rise of temperature from the graph. If necessary the graph may be extrapolated.

6. To study Blackman's law of limiting factors (interaction of environmental factors)

EXPERIMENT: Experiment is set up as in Expt. 5(*b*) keeping light intensity and temperature constant. The concentration of CO_2 is increased until the rate of photosynthesis becomes constant. Now the intensity of light is increased and the rate of photosynthesis is measured until again a constant level is reached. Then the temperature of water is increased and the rate of increase in photosynthesis is noted.

RESULTS: The results are graphically plotted as shown in the model graph (Figure 30).

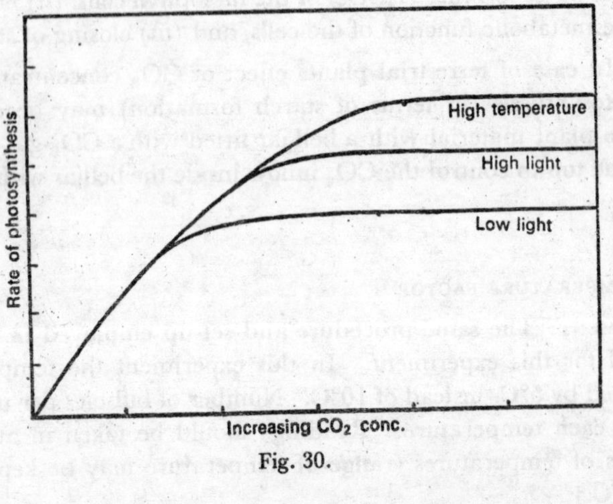

Fig. 30.

DISCUSSION: The rate of photosynthesis at a constant temperature and light intensity increases with increasing concentration of CO_2 up to a certain limit when light or temperature becomes limiting. At this stage increase in light intensity, keeping other two factors constant, again increases the rate to a point when temperature becomes a limiting factor. Increase in temperature up to a limit hastens the rate of photosynthesis.

Under a given set of conditions the rate of a process may, therefore, be largely controlled by one or two of the limiting factors and not by all the essential factors. This principle is called Blackman's law of limiting factor since he first proposed it. This is evident from the above experiment.

7. To study the methods of measuring the rate of photosynthesis

(*a*) RATE OF PHOTOSYNTHESIS IN AQUATIC PLANTS

EXPERIMENT: Rate of photosynthesis may be measured by counting number of bubbles per unit time (Refer Expt. 1.*a* or 1.*b*) or by measuring the volume of O_2 evolved per unit time (Refer Expt. 5.*b*).

N.B. The rate may also be measured by Winkler's titration procedure which gives the mg or ml of O_2 evolved (i.e., remains in dissolved condition) by constant amount of plant material per hour.

(*b*) RATE OF PHOTOSYNTHESIS IN TERRESTRIAL PLANTS WITH THE HELP OF GANONG'S PHOTOSYNTHOMETER

EXPERIMENT: Ganong's photosynthometer is a standard apparatus by which an estimation of amount of CO_2 absorbed at a given time during photosynthesis by green plants can be made. With accurate handling, the amount of O_2 evolved during photosynthesis can also be measured by this apparatus and an idea about photosynthetic quotient (O_2/CO_2) can be obtained.

The apparatus essentially consists of three parts: (a) a glass bulb fitted on a wooden base, (b) a graduated tube with a stopcock, and (c) a connecting link fitted with a stopcock which connects the graduated tube with the glass bulb. The volume of the whole set when fitted is 103 ml (Figure 31).

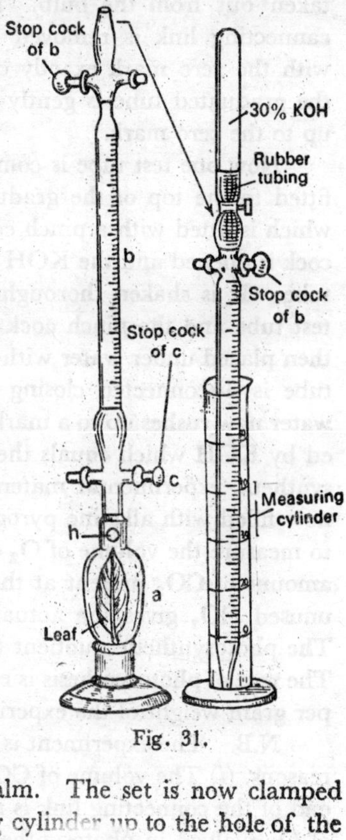

Fig. 31.

About 3 ml (measured by displacement of water) of the experimental material (any suitable fresh green leaf) is placed in the glass bulb. The volume of the apparatus now becomes 100 ml. The graduated tube is now inverted, its stopcock is closed and filled with water up to a desired mark. The graduated tube is then fitted with the connecting link. Now stopcock of the connecting link (c) is closed and its other hollow end is filled with water. It is then inverted into a suitable measuring cylinder containing water closing the hollow end with the palm. The set is now clamped keeping the level of water in the measuring cylinder up to the hole of the connecting link (h).

Now the top of the graduated tube is connected to a CO_2 generator or Kipp's apparatus. Both the stopcocks of the graduated tube and the connecting link are open and CO_2 is admitted into it. The water in the graduated tube gradually flows down and the tube is filled up by CO_2. The stopcock of the graduated tube is closed when the level of water in the graduated tube falls to the level of water outside. The tube now contains desired amount of CO_2. The whole set is now fitted at the top of the glass bulb (a), so that the hole of the connecting link coincides with that of the glass bulb. This ensures that the pressure inside the bulb and the hollow end of the connecting link is at atmospheric pressure. The connecting link is then slightly twisted to cut off the connection with the atmosphere. All connections are then made air-tight. Now the stopcock of the connecting link is opened to allow CO_2 to diffuse into the bulb. The whole apparatus is kept in bright sunlight for about three hours. After noting the time the stopcock of the connecting link is closed and the graduated tube is taken out from the bulb. It is then put in a basin of water and the cannecting link is removed under water. The graduated tube is fixed with the zero mark exactly on the surface of water and the stopcock of the graduated tube is gently opened to allow water to rise inside the tube up to the zero mark.

Now one test tube is completely filled with 30% KOH solution and is fitted to the top of the graduated tube with the help of a rubber tubing which is fitted with a pinch cock. The tube is removed from water, pinch cock is opened and the KOH solution is allowed to flow into the graduated tube. It is shaken thoroughly, KOH solution is drained back into the test tube and the pinch cock is closed. The end of the graduated tube is then placed under water with its zero mark at the surface level and the test tube is disconnected closing the stopcock of the graduated tube. The water now rushes up to a mark corresponding to the volume of CO_2 absorbed by KOH which equals the volume of CO_2 left unutilised by the photosynthetic experimental material after the desired time. The test tube is now filled with alkaline pyrogallate solution and the procedure is repeated to measure the volume of O_2 evolved. The difference between the actual amount of CO_2 present at the beginning of experiment and the amount of unused CO_2 gives the actual amount of CO_2 utilised in photosynthesis. The photosynthetic quotient (O_2/CO_2) may be calculated from the results. The rate of photosynthesis is expressed as ml of CO_2 absorbed or O_2 evolved per gram weight of the experimental material per hour.

N.B. The experiment is somewhat erroneous because of the following reasons. (*i*) The volume of CO_2 left inside the bulb as well as in the hollow end of the connecting link is neglected. (*ii*) The measurement of volume of O_2 evolved in photosynthesis is not accurate because this disregards th volume of O_2 already present in the atmosphere inside the apparatus (air contains about 16% O_2), and (*iii*) CO_2 is slightly soluble in water.

(*c*) Rate of photosynthesis in terrestrial plants by measuring dry weight increase or net assimilation

Experiment: Several weighing bottles are dried in an oven at 100°C., cooled in a desiccator and accurately weighed. These are stored in a desiccator until ready for use. Bottles are serially marked. Two suitable potted plants are kept in a dark room for a day before the experiment is started. The following morning at a convenient hour one potted plant is transferred to a sunny place and 50 discs are cut out from the leaves with a cork borer of 1 cm diameter. During cutting of discs, the leaf may be supported against a cork or rubber stoper. A representative sampling of all the leaves on the plant is taken except the very young and senescent leaves. The samples are taken in weighing bottles and fresh weights and dry weights are determined.

The plant from which the discs have been collected is allowed to remain in the sunny place for at least four hours. The duration of experiment, temperature and light intensity are recorded. The plant is adequately watered during these periods. After four hours the discs are similarly cut out, taken in weighing bottles, fresh and dry weights are determined. The experiment is repeated at an interval of four hours as long as time permits.

The same procedure is repeated in case of plants kept in dark.

Results: Increase in dry weight per four hours by 50 discs is calculated in each case and the result is expressed as mg dry weight increase per sq. cm of the leaf per hour.

N.B. The amount of CO_2 consumed per hour per sq. cm may be calculated, assuming all the photosynthate as glucose, as follows.

$$\frac{44}{180} \times \text{Increase in dry wt. in mg/hr/sq. cm}$$

where 180 = Mol. wt. of glucose,
44 = Mol. wt. of CO_2.

Again the number of litres of air required to supply the CO_2 absorbed per sq. cm per hour may also be calculated as follows. If amount of CO_2 consumed/hr/sq. cm of leaf is taken as X, volume of CO_2 in air as 0·03 per cent and weight of 1 ml of CO_2 as 1·977 mg at standard conditions then the amount of air is $\frac{X \times 100}{0{\cdot}03 \times 1{\cdot}977}$ ml of air.

8. Determination of real and apparent photosynthesis

EXPERIMENT: Two experimental set-ups are arranged as in Expt. 2.*a* (*iv*). Here tap water is used (which may be previously oxygenated by photosynthesis). The experimental materials should be of comparable number and weight in each case. The initial O_2 content of water in each set-up is determined following the procedure given in the referred experiment.

One set-up is kept in bright light for photosynthesis and the other in dark for respiration only. After two hours the O_2 content of water of both the set-ups is again determined.

RESULTS: The difference between the final and the initial content of O_2 in case of the set-up kept in light gives the excess of O_2 liberated by photosynthesis. This amount indicates apparent photosynthesis. Again the difference between the initial and final content of O_2 in case of the set-up kept in dark gives the amount of O_2 consumed during respiration. When this amount of O_2 is added to the amount of O_2 liberated by photosynthesis it gives the real photosynthesis.

DISCUSSION: When the net result of photosynthesis is measured and no correction is made for respiration it is called the *apparent photosynthesis*. When, on the other hand, corrections for respiration are made, the photosynthesis thus calculated is called the *real photosynthesis* or *true photosynthesis*. So to obtain a measure of true photosynthesis it is desirable to make corrections for respiration. Some of the O_2 given off in photosynthesis is used up in respiration and some of the CO_2 given off in respiration is taken up in photosynthesis. It is therefore necessary to measure the respiration of an identical sample in dark. It is easily seen that the apparent photosynthesis is less than the true photosynthesis by the amount of O_2 used up in respiration.

The respiration error is variable and is always present, and it thus complicates attempts to measure true photosynthesis, since the factors which influence photosynthesis are likely also to influence respiration.

9. To show that the entry of CO_2 takes place through stomatal pores

EXPERIMENT: A suitable potted plant is so selected that its leaves are dorsiventral and upper surfaces are thickly cuticularised. The plant is kept in dark for two days to deplete their leaves of starch. Now some leaves are smeared with vaseline only on the lower surfaces and other leaves are smeared only on the upper surfaces. Vaseline should be applied as a thin film so that stomatal pores are closed. The plant is then kept in light for several hours. After that period the leaves are severed and tested for starch by iodine.

OBSERVATION: It is observed that the leaves having vaseline applied on the lower surface only do not show the presence of starch whereas the leaves in which vaseline was applied on the upper surface shows the presence of starch by iodine.

INFERENCE: In case of the leaves smeared with vaseline on the lower surface starch synthesis cannot take place because CO_2 could not enter into the leaf for use in photosynthesis owing to the closure of stomatal pores by vaseline. Those leaves in which vaseline has been applied on the upper surface have their stomata open on lower surface through which CO_2 could enter and allow the photosynthesis to occur resulting in starch synthesis. Thus the experiment shows that the CO_2 enters the leaves through stomatal pores of the lower surface in case of dorsiventral leaves.

10. To show the necessity of light for synthesis of chlorophyll in leaves

EXPERIMENT: Some pea or gram seeds are allowed to germinate in dark as well as in light.

OBSERVATION: After 10 days the dark grown seedlings show yellow or white leaves and stems. Other seeds grown in light show normal seedlings with green leaves and stems. When etiolated plants are exposed to light these turn to green.

INFERENCE: The experiment shows that light is necessary for chlorophyll synthesis. In a light-mediated reaction protochlorophyll is reduced to form chlorophyll a. Photoreduction of protochlorophyllide to chlorophyllide a, followed by phytol esterification to form chlorophyll a, is now thought to be the major pathway.

11. To show that starch synthesis in etiolated (or albino) leaves are independent of light provided sugars are present

EXPERIMENT: Two comparable leaf samples are picked from etiolated rice or wheat seedlings grown in a dark room. Each sample is placed in a beaker of 100 to 150 ml capacity and some weight is placed on the leaves to prevent them from floating. One sample is covered with pure water and the other with 0·1 M sucrose solution for two days in the dark room. The samples are removed from the beaker, excess liquid is blotted off, placed on moistened filter paper in petridishes and set aside in the dark room for a period of 24 hours. At the end of the experimental period the leaves from both the sets are tested for starch by iodine.

OBSERVATION: No starch is obtained in case of leaves kept in water whereas starch synthesis takes place in case of leaves kept in sucrose solution.

INFERENCE: The experiment shows that the starch synthesis in etiolated leaves is independent of light if sugars are supplied to the etiolated leaves.

CHAPTER XII

GROWTH AND GROWTH REGULATORS

Growth may be defined as a permanent and irreversible change in volume or size of an organism accompanied by an increase in dry weight. The term is employed to designate the complex processes of the continuous increase in size of plants and production of new organs at least during some stage of the life period. This growth phenomenon is dependent upon physiological processes. Besides, other factors specific to growth and development such as *plant hormones*, *vitamins*, *inhibitors* and other *growth-regulating substances* must be mentioned.

The complete development of a plant cell or of an organ may be seen to occur in three distinct although overlapping steps. The three steps which occur in meristems are *cell division*, *cell enlargement* and *cell differentiation.* The growth of a plant may be due either to cell division or to cell enlargement. Thus the growth of a plant or plant organ is essentially an increase in size due to enlargement of cells after they divide and this cell enlargement must be irreversible. This is not a reversible increase like that of increase due to hydrostatic pressure caused by turgour. The cell division and enlargement cannot long continue in plants without a minimum of differentiation, but the course of development may be very different.

The mechanism of growth of a cell involves the following processes: (*i*) absorption of water, (*ii*) absorption of solutes for maintaining higher turgour pressure of the cells, (*iii*) deposition of new cell wall materials on the stressed cell walls (by intussusception and by apposition), (*iv*) formation of new protoplasm by synthesis of new proteins, lipids, etc. for the synthetic processes.

The whole growth sequence can be divided into four phases from beginning to end and can be graphically represented or illustrated as a *sigmoid growth curve* consisting of following phases: (*i*) initial lag phase (slow rate) of growth, (*ii*) log phase of growth (the period of maximum growth rate) and is called *grand period of growth*, (*iii*) declining phase of growth, (*iv*) senescent phase of growth.

There are certain factors in the external physical and chemical environment of a plant and certain other factors produced within the plant body (growth promoters and growth inhibitors), which control and regulate the pattern of growth and development of plant. Every plant which grows and develops must obtain certain special types of nutrient materials in suitable amounts at various stages. These are: (*i*) Carbonic food materials, e.g., carbohydrates, fats, proteins. Chlorophyll is also an essential substance in autotropic plants. (*ii*) Certain carbonic compounds which occur in minute quantities within the plant having important and essential role, such as enzymes and vitamins. Besides, the hormones are the organic compounds which regulate the growth and development of plant organs. (*iii*) Inorganic mineral elements and salts which are essential in influencing the growth directly or indirectly. (*iv*) Water is the most important factor, without which growth is impossible. (*v*) Oxygen which is used in respiration. (*vi*) CO_2 which is required for photosynthesis.

Among substances which markedly influence reactions and metabolism of plants are the growth substances which include three well known groups, such as (*i*) auxins, (*ii*) cytokinins, and (*iii*) gibberellins. These are produced within the plant body in minute quantities and exert their effects on cell division, cell enlargement and cell differentiation. Nowadays ethylene must be included in the list of growth regulating substances since a number of important physiological processes have been found to be controlled by ethylene and most of the actions of auxin are thought to be mediated through ethylene. Hence it is often called a 'gaseous hormone'.

Another group of substances which markedly influence growth and development of plants are inhibitors. The inhibitors include a wide array of chemical entities which may inhibit growth or developmental functions or may inhibit some enzyme-dependent reaction or some component reaction relating to the growth regulators. The natural inhibitors of plants are utilized to control or limit growth processes, but the presence of toxic substances or inhibitors in plant extracts is not necessarily indicative of their involvement in any growth or developmental process. The common naturally occurring inhibitors are abscisic acid, ferulic acid, coumaric acid, caffeic acid, chlorogenic acid, and other phenolics. They are mostly inhibitors of auxins though in low concentrations they may have a synergistic effect on growth. Some synthetic compounds like CCC (cycocel), Amo-1618, Phosphon-D, etc., are known as antigibberellins, as they antagonise gibberellin action and also prevent gibberellin biosynthesis. Apart from the naturally occurring inhibitors of auxin, there are some synthetic inhibitors like TIBA (triiodobenzoic acid), 2, 6-D (2, 6-dichlorophenoxy acetic acid), 2, 4-isobutyric acid, etc., which competitively inhibit auxin activity.

Besides these, there are some inhibitors of metabolic processes or enzyme systems and information on these inhibitors is widely distributed in scientific literature. Application of poisons and inhibitors on biological systems reveals much empirical knowledge concerning the effects of a wide variety of inorganic substances (e.g., fluoride, cyanide, arsenite, hydrogen sulphide, salts of mercury, lead, iron, barium, etc.) and of organic materials (e.g., antiseptics and narcotics such as toluene, chloroform, chloral, dye-stuffs, etc.) on enzyme preparations and on cell life in general. The respiratory inhibitors CO and cyanide provided much valuable early knowledge in the development of this field.

Sometimes a number of compounds have the ability to combine with certain enzymes, yet they do not serve as substrate and, therefore, block catalysis by that enzyme. These compounds are called competitive and noncompetitive inhibitors. When a compound competes with a substrate or coenzyme for the active site on the enzyme protein and thereby reduces the catalytic activity of that enzyme, the compound is considered to be a *competitive inhibitor*, e.g., malonic acid. This inhibition can be reversed by increasing in turn the concentration of the substrate, i.e., succinic acid. The type of inhibition that cannot be reversed by increasing substrate concentration is called *noncompetitive inhibition*, e.g., reaction of iodoacetamide on triosephosphate dehydrogenase. The inhibition of many enzymes by heavy metal ions such as Ag^+, Hg^{++}, Pb^{++}, etc. is also example of non-competitive inhibition.

The utilisation of the required metabolites by the organism is dependent on enzymes, and these enzymes may be inhibited by substances whose structures are related to those of the metabolites. Although in most cases, the nature of the essential enzymic step is unknown, the term *antimetabolites* is used to describe these inhibitors. Antimetabolites may also block the utilisation of metabolites that are synthesised by the organism itself. The antimetabolites that have been obtained from living organisms, such as bacteria, fungi, actinomycetes, etc. are commonly called antibiotics. There are several purine and pyrimidine analogues and a number of antibiotics which function as antimetabolites.

There are also some chemical compounds which act as uncoupler of phosphorylation, i.e., uncoupler of oxidative phosphorylation and inhibitors of photophosphorylation, e.g., DNP, DCMU, CMU, DCIP, etc.

Some commonly used enzyme inhibitors

Inhibitors	Function
Cyanide	Inhibits enzymes containing Fe, Cu, Zn.
Sulphide	Inhibits enzymes containing various metals.
Fluoride	Inhibits enzymes containing Mg, Ca and other metals.
Oxalate	Inhibits enzymes containing Cu, Mg.
CO (Carbon monoxide) ...	Inhibits enzymes containing Fe, Cu.

Inhibitors	Function
Diethyldithiocarbamate ...	Inhibits enzymes containing Cu.
Azide	Inhibits enzymes containing Fe-protoporphyrin enzymes.
EDTA (Ethylenediamine tetra-acetic acid)	Inhibits enzymes containing various metal ions.
Heavy metals like Ag, Hg, Pb, etc.	Inhibits enzymes containing sulphydryl (may also cause non-specific protein precipitation)
Iodoacetate	Inhibits enzymes containing sulphydryl, imidazole, carbonyl, thiolester.
Various arsenicals	Inhibits enzymes containing sulphydryl.
5-Fluorodeoxyuridine ...	Specific inhibitor of DNA in plant tissues.
Mitomycin-C	Inhibits the synthesis of DNA. DNA replication is also prevented.
Aotinomycin-D	Inhibits DNA-dependent RNA synthesis by RNA polymerase. RNA chain elongation and, not initiation, is blocked.
Rifamycin, Rifampicin ...	Inhibitors of RNA polymerase. Inhibits synthesis of all cellular RNA while DNA synthesis remains unaffected. RNA polymerase of nuclear origin in eucaryotic cells is resistant to even very high concentration.
Cycloheximide (Actidion) ...	Major effect on t-RNA methionine binding, t-RNA release and on m-RNA-Ribosome movement.
Chloramphenicol	Binds specifically to 50S subunits. Inhibits functional attachment of the aminoacyl end of AA-t-RNA to the 50S subunits thus inhibiting transpeptidation.
2-thiophene alamine ...	Phenylalamine analogue.
3-Fluorotyrosine	Tyrosine analogue.
5-Flurotryptophane	Tryptophane analogue.
Ethionine	Methionine analogue.
D-Glucosamine	D-Glucose and D-Fructose analogues.

Inhibitors	Function
6-Mercaptapurine, 6-Methylpurine ...	... Purine analogue.
8-Azaguanin ...	... Purine analogue.
2-Thiouracil ...	... Pyrimidine analogue.
6-Azacytosin ...	... Pyrimidine analogue.

A. EXPERIMENTS ON GROWTH

1. Study of meristems

EXPERIMENT: Medium longitudinal microtomic sections are cut through the stem tip and root tip of some suitable herbaceous species and are observed under high power of microscope.

OBSERVATION: The apical meristematic region, embryonic leaves, bud primordia, nodes and internodes are distinguished in case of stem tip and in root tip approximate limits of the regions of *cell divisions*, *cell enlargement* and *cell maturation* are distinguished (Figure 32). Meristematic cells are usually

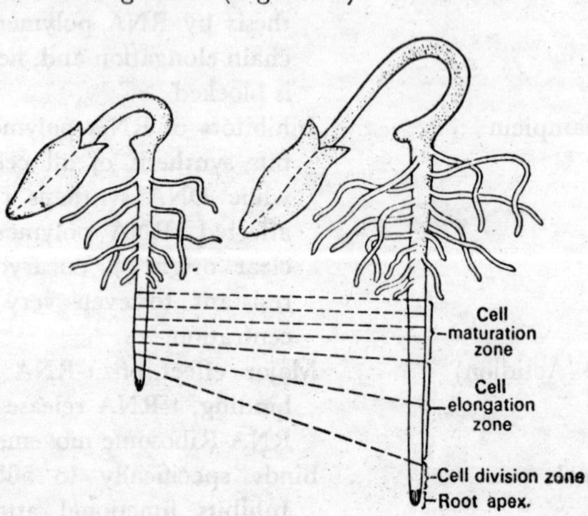

Fig. 32.

isodiametric in shape, compactly set without evident intercellular spaces, have dense cytoplasm with small vacuoles and large prominent nuclei. Plastids are in proplastid stage, cell wall thin and homogeneous. A gradual change in shape and size is observed in the derivatives which ultimately form permanent cells.

INFERENCE: Cells responsible for growth can thus be distinguished in medium longitudinal sections of apical regions of roots and stems.

2. Determination of regions of cell enlargement in leaves, stems and roots

EXPERIMENT: One of the younger leaves of a rapidly developing suitable plant is selected. Now with the help of a scale two sets of lines at right angles to each other are drawn with India ink so that the leaf is marked with squares of equal sizes (marking can also be done with the help of space marker disc).

In another set a suitable plant which has developed sufficiently to have at least two to three internodes is selected. The stem portions of the upper three internodes are marked with India ink with short horizontal lines spaced equally apart.

In a third set some seeds of suitable species are allowed to germinate in moist filter paper. One of the seedlings with a root about 2 cm length is selected and the entire root is marked with India ink with horizontal lines spaced equally apart and seedling is allowed to grow in a bottle (Figure 33).

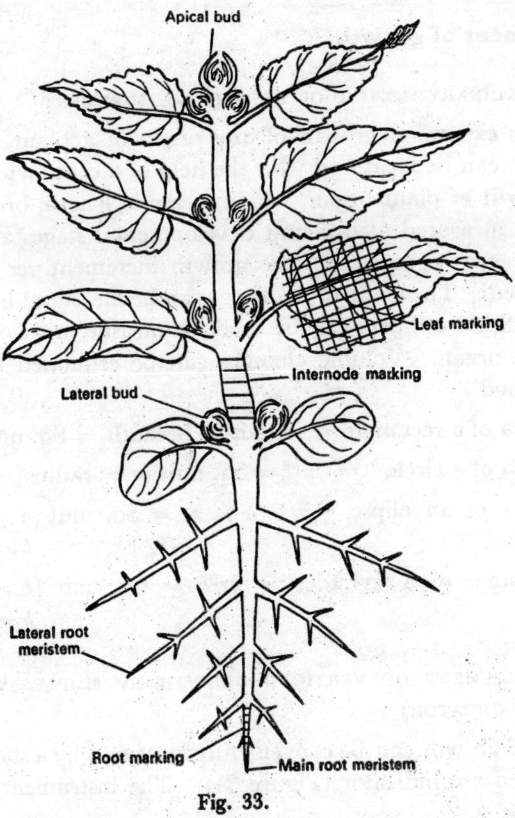

Fig. 33.

OBSERVATION: After about three days the leaf, stem and the root are remeasured considering the area of the squares in case of the leaf and distances between two horizontal lines in case of stem and root.

INFERENCE: Measurement shows that maximum increase in area of the squares takes place along the edge of the leaves which indicate that meristems (*plate meristem*) are located along this region. In case of stem maximum growth is found just above the nodes. This growth is due to *intercalary meristem* left behind by the apical meristem which go ahead during development. In case of root maximum growth is found just below the apical region which is due to *apical meristem*. It is observed that the distance between the first and the second line near the tip has not separated much by growth, and this includes the *zone of cell division* by apical meristem. But the distances between the lines behind the meristem zone increases maximum amount due to growth by elongation and this includes the *zone of cell elongation*. Still further behind this zone the distance between the lines again is not marked and this zone includes the *cell maturation zone*.

3. Measurement of growth

(a) AN ORDINARY METHOD OF GROWTH MEASUREMENT

The linear expansion growth of any organ of a plant, i.e., length or breadth or area can be measured with the help of a centimeter scale. The expansion growth of plant organ, i.e., increase in length, breadth, area or volume can be measured at different developmental stages and at various intervals. The growth rate, i.e., the growth increment per unit time can also be calculated. The elongation of root can be measured by any scale or graph paper. Slide cellipers can be used for measurement of thickness or diameter of an organ. Volume changes can be estimated by water displacement method.

N.B. Area of a rectangle = length × breadth = Sq. unit.

Area of a circle = πr^2 = Sq. unit (r = radius; π=3·142).

Area of an elipse =$\pi/4\ (a \times b)$ = Sq. unit (a = long axis, b = small axis).

Volume of a cylinder = $\pi r^2 h$ = Cu. unit (r = radius, h = height).

(b) MEASUREMENT OF VERTICAL GROWTH BY SIMPLE AUXANOMETER (ARC INDICATOR)

The rate of growth can be conveniently measured by a standard simple instrument called arc indicator (Figure 34). The instrument consists of a metallic sextant scale (in angular degrees) fitted to a vertical stand by means of arms which meet forming a right angle with each other. At this end they are fixed to an axle fitted with a pulley. The axle holds a pointer which can move up and down on the sextant scale when the pulley rotates. A thread is wound on the groove of the pulley. One end of this thread is tied at the apex of a growing plant while from the other end hangs a suitable weight.

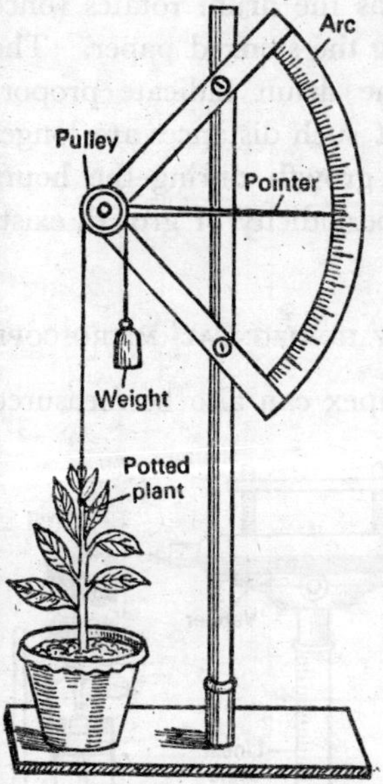

Fig. 34.

As the plant grows in height the pulley rotates by the downward movement of the weight. With the rotation of the axle the pointer moves on the sextant scale and this movement is proportional to the growth in height of the plant. By recording the movement of the pointer by angular degrees at particular intervals the growth rate of the plant can be measured. One angular degree $= \frac{\pi}{180} = \frac{3{\cdot}142}{180}$.

In linear measurement one angular degree $= \frac{3{\cdot}142}{180} \times$ radius of the arc (cm).

(*c*) Measurement of Vertical Growth by Drum Auxanometer

This is a more complex instrument and gives more detailed idea of growth rates both during day and night (Figure 35). It consists of a revolving drum with a smoked paper pasted on it. The apical region of a growing potted plant is tied with one end of a thread which passes over a pulley holding a weight at the other end. The pulley is fixed to a vertical stand: Another thread passing over a bigger pulley holds a weight at one end while to the other end, a

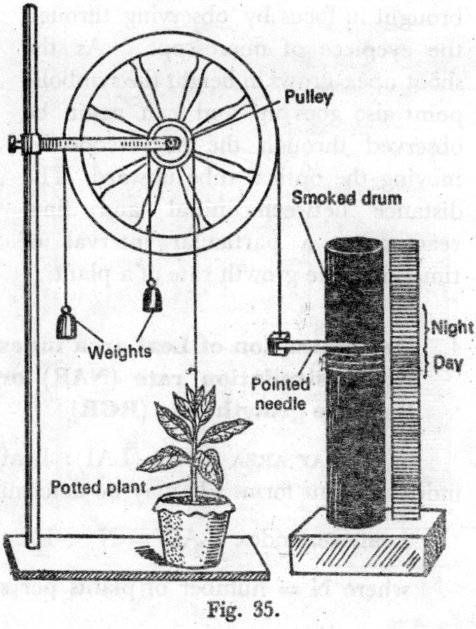

Fig. 35.

pointer is fixed horizontally in such a way that only the sharp end of the pointer just touches the surface of smoked paper and its course of movement can be traced on it. As the shoot apex grows in height the hanging weight pulls the pointer upward and as the drum rotates (once in every hour), the pointer traces its path along the smoked paper. The perpendicular distances between tracing on the drum indicate proportional hourly growth. It should be noted that such distances are longer at night compared to day indicating increased growth during the hours of darkness. The result also indicates that a periodicity of growth exists in plants.

(*d*) MEASUREMENT OF VERTICAL GROWTH BY HORIZONTAL MICROSCOPE

The vertical increase in height of a shoot apex can also be measured with the help of an optical microscope. The optical tube of this microscope remains in horizontal position (Figure 36). The optical tube can be moved upward and downward with the help of a screw and the vertical movement can be measured with the help of a linear scale attached to the stand. Now a small point is made with ink on the shoot apex which is brought in focus by observing through the eyepiece of microscope. As the shoot apex grows in height the symbolic point also goes up and can again be observed through the microscope by moving the optical tube upward. The distance between initial and final reading at a particular interval of time gives the growth rate of a plant.

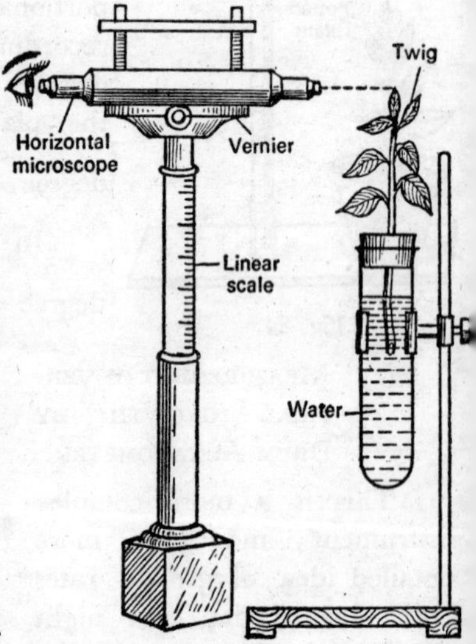

Fig. 36.

4. Determination of Leaf area index (LAI); Leaf area ratio (LAR); Net assimilation rate (NAR) or Unit leaf rate (ULR); and Relative growth rate (RGR)

(*i*) LEAF AREA INDEX (LAI): Leaf area index is the morphological index of plant form. It may be determined in the following way:

Leaf area index (LAI) $= N \times L_A$.

where N = number of plants per square metre of the ground area and hence the area of the ground per plant is $\frac{1}{N}$. The leaf area divided by this area $\left(\frac{1}{N}\right)$ gives the LAI which is a pure number.

LA can be determined in field condition by constant multiplying method as follows. Actual areas of different leaves of the plant are determined individually by graph paper method (Refer Expt. 5.*c* of chapter VI) and average of them is calculated. Let it be *a*. Again areas of these leaves are individually determined by multiplying their length and breadth and average of them is calculated. Let it be *b*. Now *a*/*b* may be taken as a constant (K). Thus to get the area of any leaf of a plant roughly, its length is first multiplied by its breadth and then with the constant K.

(*ii*) LEAF AREA RATIO (LAR): Leaf area ratio gives an indication of the dry matter accumulation by the leaf in relation to its area. It may be determined in the following way:

$$\text{Leaf area ratio (LAR)} = \frac{L_A}{L_W} \text{ cm}^2/\text{gm}$$

where LA = Area of a leaf

LW = Total dry weight of the leaf.

(*iii*) NET ASSIMILATION RATE (NAR) OR UNIT LEAF RATE (ULR): This is a physiological index (rate of increase of dry weight of the whole plant per unit leaf area) closely connected with the photosynthetic activity of the leaves. It relates dry weight increase, which in most plants has carbon assimilation as its most important components, to the area of those organs most concerned with carbon assimilation, i.e., leaves. It may be determined in the following way:

$$\text{Net assimilation rate (NAR)} = \frac{W_2 - W_1}{T_2 - T_1} \times \frac{\log_e (L_2 - L_1)}{L_2 - L_1} \text{ gm/cm}^2\text{/day}$$

where W_1 = Dry weight (gm) of the plant at time T_1 (day),

W_2 = Dry weight of the plant at time T_2,

L_1 = Leaf area (sq. cm) at time T_1,

L_2 = Leaf area at time T_2,

$\log_e$ = 2·33.

(*iv*) RELATIVE GROWTH RATE (RGR): This indicates the overall growth index, i.e., the growth taking place in the plant daily and may be determined as follows:

$$\text{Relative growth rate (RGR)} = \frac{\log_e (W_2 - W_1)}{T_2 - T_1} \text{ gm/plant/day.}$$

5. Demonstration of seasonal periodicity in growth of woody perennials

Experiment: This experiment should be performed in four different seasons of the year, viz., spring season (February to April), summer season (May to July), rainy season (August to October) and winter season (November to January). At the beginning of the spring season several branches on a young tree or shrub are selected and buds on each are tagged. The diameter of each branch at the base is measured. The length of each developing bud is also measured. This measurement is continued at an interval of 10 days throughout four seasons. The average of the measurement of each season is recorded.

Results: The results are plotted taking average increase in growth per 10 days in each season as an abscissa and days (90) in each season as ordinate.

Discussion: The experiment shows that there exists a seasonal variation in growth pattern in plants. The maximum increase in growth takes place during the growing season, i.e., the spring and this decreases to a minimum during the lean months of the winter season.

6. Demonstration of growth correlation

Experiment: Three uniform sized leaves of *Bryophyllum* are taken. Now narrow transverse strips are cut from one leaf through the midrib allowing the leaf to remain attached near the margins. Again narrow transverse strips are cut from the second leaf allowing the midrib to remain intact. The third leaf is also cut transversely into a number of segments through the lobes and not through the notches (Figure 37).

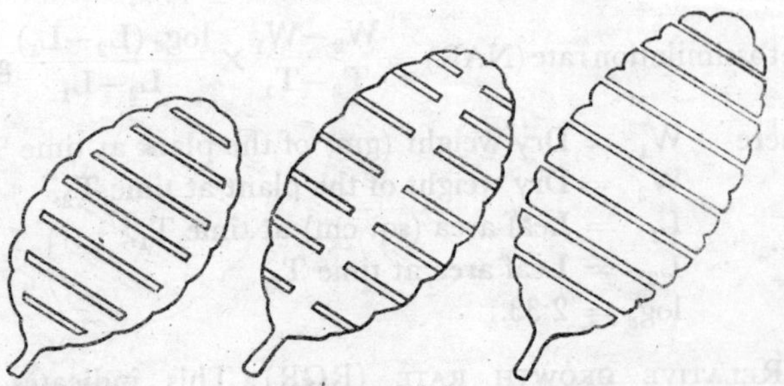

Fig. 37.

Each leaf is placed on moist filter paper in a covered petridish and kept out of direct sunlight. One control leaf is also placed in another petridish. Filter papers are remoistened from time to time.

RESULTS: After four weeks the number, height and position of new plants growing from the notches in the margin of the leaves are recorded and compared.

DISCUSSION: The abilities of one part of the plant to direct growth and development in another is called correlation effects. Tip of the leaf is necessary for perception of stimulus but not for the response which takes place further down the tip. In the present experiment the tip is completely separated from the other parts in third leaf. In second leaf the notches are separated where the mid-vein is the connecting link between different parts of leaf. Whereas in case of first leaf the notches are directly connected with the tip. It is observed that maximum number of normal plants (less than control) appear in case of the first leaf, then in second and minimum in third.

7. To show the root/shoot ratio at various experimental conditions

EXPERIMENT: Four pots containing washed sand are taken. Ten seedlings (*Vigna* or *Phaseolus*) of uniform size and age are transplanted in each pot. One pot is placed in bright light, the second in shade, the third is treated with 10 ml of calcium nitrate (concentration 0·820 mg/litre $Ca(NO_3)_2$) solution and the fourth is treated with 1 ml of this calcium nitrate solution.

RESULTS: The growth of plants in each pot is observed over a period of 10 to 15 days. At the termination of the experiment the shoot and root portions are separated and fresh weights are taken after cutting the shoots and roots into small pieces. The shoots and roots are then dried at 100°C. in an oven for 72 hours and these are reweighed to determine the dry weight of each. The ratio of root to shoot for each treatment is calculated on both fresh and dry weight basis and results are tabulated.

DISCUSSION: The root by shoot ratio is higher in plants grown in light compared to plants grown in shade. Again the root by shoot ratio is higher in plants grown in low nitrogen level compared to those grown in high nitrogen level. In fact the effect of a given treatment is likely to favour root growth relative to that of shoot or vice-versa, is decided by the concentration of various other essential factors such as organic food, light, hormones, minerals, oxygen, water content, etc., which often limit to a greater degree the process of growth.

8. Demonstration of grand period of growth

EXPERIMENT: This experiment may be performed taking growth rate of root tip or of any suitable plant, or unicellular organism.

A fair number of plants grown under favourable conditions in open atmosphere may be used for taking growth data. Increase in height of the plant is recorded at an interval of seven days from germination until increase in height ceases.

RESULTS: The recorded results are graphically plotted taking time (weeks) as abscissa and increase in length or height as ordinate.

DISCUSSION: In case of root growth it is observed that the root tip first grows slowly then rapidly, and finally slowly until it stops elongation. Such changes constitute the grand period of growth and a curve resembling that obtained for a single root may be obtained for an entire plant. In case of shoot growth the rate of growth gives a similar rise and fall, whereas total height of plant of course does not drop. The curve (Figure 38) will be of "S"-shape (sigmoid curve). The growth of the entire plant or of the plant organ characteristically passes through the stages represented by this curve. The time during which this occurs is called the grand period of growth.

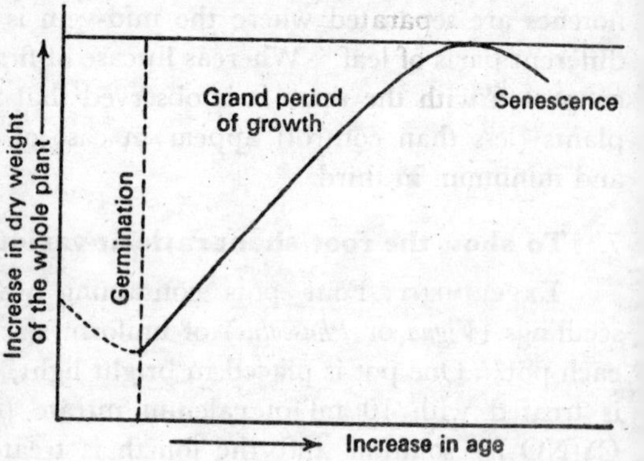

Fig. 38.

9. Demonstration of the effect of water supply on growth of plant

EXPERIMENT: Four suitable potted plants are taken and are allowed to become dry so that the plants just start to wilt. One pot is watered to optimum, second to maximum, third to minimum and the fourth is kept as control (not watered). The growth rate is measured with the help of auxanometer.

RESULTS: The growth rates of each potted plant is separately taken and graphically represented.

DISCUSSION: Maximum growth is observed in optimum watered plant which is followed by maximum watered, minimum watered and the plant of the fourth pot, which has not been watered at all, wilts. Water is essential for normal growth of all plants. All living tissues contain some water, and more active tissues such as leaves, growing roots and stems rarely contain less than 50% water. Growth inhibitions are induced by wilting due to closure of stomata and inhibition of photosynthesis. Heavy watering of the pots interferes with the aeration of roots due to water logging. Hence growth rate is inhibited.

N.B. This experiment may also be performed with fruiting tomato plants or other fleshy-fruited plants if available in pots. The effect of water upon the rate of growth of fruits (measured in terms of diameter of fruits) may be observed.

10. Demonstration of effect of light on plant growth

EXPERIMENT: Two suitable well-watered potted plants are selected. One pot is kept under bright sunshine (or under artificial light—200 W bulb fitted with a reflector at a distance of 3 to 4 feet from the plant). The other is kept in shade. The growth rates of the plants are measured with the help of an auxanometer.

RESULTS: The growth rates are graphically plotted.

DISCUSSION: The growth rates will be higher in plant kept in light compared to the plant kept in shady place. The rate of growth is higher under optimum light intensity whereas the effect of high light intensity is in general retarding on growth process. The effect of light is mainly through photosynthesis and transpiration. The effect is also interrelated with the effect of temperature and water.

N.B. Dark-grown seedlings sometimes show higher growth rate (in height) compared to light-grown ones. However, dry weight increase is more in case of the latter.

11. Demonstration of the effect of temperature on plant growth

EXPERIMENT: This experiment can be performed by recording the growth rates of roots or shoots kept under different temperatures.

Two suitably grown well watered plants are taken. One is kept at a temperature of 20°C. and the other at 30°C. under well illuminated condition. The growth rate is measured at an interval of 7 days for 4 to 5 weeks in terms of dry weight and fresh weight of the plants.

RESULTS: The growth rates of the plant under two different temperatures are graphically plotted.

DISCUSSION: The growth rate increases with increase in temperature within a particular range (25 to 35°C.). Both high and low temperatures inhibit growth. The effect of temperature on growth is generally controlled by enzyme activity. Each of the physiological processes is directly or indirectly influenced by varying degrees of temperature.

12. Demonstration of effect of oxygen on plant growth

EXPERIMENT: Two 100 ml rubber-stoppered bottles are taken. To one bottle 2 gms of pyrogallic acid and 10 gms of KOH in 60 ml of water are added. The bottle is shaken well. The oxygen within the bottle will be absorbed by it. The alkaline pyrogallate solution is thrown off and stoppered carefully. The radicles of two or three germinating seedlings are marked with India ink and kept in two petridishes containing water. The oxygen free bottle is held inverted covering the seedlings of one petridish. The seedlings of the other petridish are similarly covered with the other bottle. The growth rate is measured in each case.

RESULTS: The changes in growth rates are graphically plotted in each case and compared.

DISCUSSION: The rate of growth of radicle will be higher in seedlings covered with bottle containing normal air compared to the bottle containing oxygen-free air. Oxygen in molecular form plays an important role in growth phenomena chiefly through the influence on respiration. Usually a depletion of oxygen is associated with an increase in carbon dioxide and this may further depress the root growth as well as salt and water absorption. In plant growth porosity of soil is of major importance in influencing the aeration and, therefore, the oxygen supply to roots.

13. Demonstration of effect of food supply on growth

EXPERIMENT: Some rice and bean seeds are separately grown. The seedlings are taken when the roots are 2 to 3 cm long. Root tips are marked with India ink as in the previous experiment. Half of the endosperm tissue from some rice seedlings and half of the cotyledonary tissue from bean seedlings are removed with the help of a sharp knife. The remaining seedlings with intact endosperms and cotyledons are kept as control. The mutilated and control seedlings are kept together in petridishes under favourable conditions.

RESULTS: The growth rates of roots of mutilated and control seedlings are measured in centimeters over a period of four days and graphically represented.

DISCUSSION: The endosperm in case of rice and cotyledons in case of beans are the reservoir of food materials which nourish the growing seedlings up to certain stage of development after which they can independently synthesise their food by photosynthesis with developing leaves. If the food supply is partly reduced by removing a portion of endosperm or cotyledon before the plant becomes independent the growth is hampered due to paucity of food supply.

N.B. This experiment may also be performed by growing half of the independently growing seedlings in dark and half in light. The growth rate in terms of fresh and dry weights may be measured and compared.

14. Demonstration of influence of fruiting upon vegetative growth of plant

EXPERIMENT: Some suitable fruit plants, e.g., tomato, cotton, etc. are grown in field plots under natural environment. The growth data of the plant, e.g., height and branch length are taken when the flower buds just appear. Flowers and young fruits are regularly removed from half of the plants as soon as they appear. When the fruits have developed to a certain degree of maturity in the other half of plants the final height and branch length of all the plants are recorded.

RESULTS: The differences in vegetative growth between the plants on which fruiting occurs and those on which fruiting is prevented are graphically (histogram) plotted and compared.

DISCUSSION: The vegetative growth of the plant almost ceases when flowers and fruits appear on it. Flowers and fruits are active mobilisation centres which draw the metabolites from other parts of the plants leading to the reduction of vegetative growth. The removal of flowers and fruits from the plants removes the mobilisation centres keeping the plant in vegetative condition.

15. Demonstration of the effect of rest period on growth

EXPERIMENT: One lot of freshly collected potato tubers and one lot of one year old potato tubers are taken for this experiment. The tubers are cut into pieces keeping the eyes intact. These are then allowed to grow in pots.

OBSERVATION: It is observed that freshly harvested potato tubers fail to sprout whereas one-year-old tubers sprout normally.

INFERENCE: The rest period is attributed to some condition within the plants or its structures. Perfectly viable seeds or buds having rest periods are characterised by the failure of germination or sprouting even when placed under environmental conditions that are ordinarily favourable for quick germination and vigorous seedling growth. As soon as the rest period is over such seeds or buds readily germinate or sprout. The possible reasons for rest period are unfavourable environment, inhibitor promoter level, defficiency of nitrogen, photoperiod, concentration of auxin and ethylene, etc.

B. EXPERIMENTS ON GROWTH REGULATORS

16. Demonstration of the effect of auxin on growth

(a) ON ROOT GROWTH AND STEM GROWTH

EXPERIMENT: 0·25% lanolin paste (wool fat) of auxin, viz., indole acetic acid (IAA) or indole butyric acid (IBA) or naphthalene acetic acid (NAA) is prepared by mixing 1 mg of auxin to 400 mg of lanolin (see also Appendix II). A small quantity of this paste is applied to one side of some young stems of sunflower, tomato or tobacco or other suitable plants and to some other plants just at the base of the stem in the form of a girdle or ring. After a few days observations are made.

OBSERVATION: Bending of stem towards the untreated region of the stem is observed. In case of plants where an auxin girdle has been made root initials or even rootlets have been found to come out.

INFERENCE: Auxin increases markedly the plasticity of the walls of elongating cells and thus reviving the stage of elongation. Greater elongation of the cells occurs at the auxin treated region of the stem than the untreated region causing the bending of the stem. It is the property of auxin to stimulate the initiation of cell division and rate of formation of roots when applied in lanolin paste to the stem. Auxin leads to remeristemisation of permanent cortical tissues and organ differentiation leading to root formation.

(*b*) ON PETIOLAR ABSCISSION

EXPERIMENT: Leaves of some suitable plants (*Ervatamia*, *Coleus*, *Phaseolus*, etc.) are debladed, i.e., lamina is removed from the petiole and at the cut end of the petioles auxin paste (0·25%) is applied keeping a few debladed untreated plants as control. Observation is made after a week.

OBSERVATION: It is observed that petioles of the untreated plants abscised while the treated petioles do not (slight but uniform pressure is to be applied to ensure abscission of petioles).

INFERENCE: It is the property of auxin to inhibit abscission. Some of the important reasons are petiolar growth, higher auxin gradient at the petiolar end and prevention of mobilisation of metabolites.

(*c*) ON APICAL DOMINANCE

EXPERIMENT: It is a common finding that when terminal bud of a shoot is excised, one or more lateral buds promptly develop below. The nature of the influence exerted by the terminal bud on suppressing the growth of the lateral buds is called *apical dominance.* Potted sunflower or *Coleus* plants are taken for this experiment and from some plants the apical buds are removed keeping other plants as control. To half of the excised plants auxin in lanolin paste (0·25%) is applied. Observation is made after 15 to 20 days.

OBSERVATION: Lateral buds come out in plants only where the terminal buds have been removed. Whereas few or no lateral buds develop in auxin-treated and in control plants.

INFERENCE: The terminal bud inhibits or suppresses the growth of lateral buds thus establishing dominance over the remaining buds which remain inactive. As soon as the terminal bud is excised the dominance of the terminal bud is removed causing the hitherto suppressed lateral buds to come out. The effect is thought to be due to the inhibitory property of auxin produced at the terminal bud which is transported downward through the stem. This is further proved by the substitution of the terminal bud with auxin (in lanolin paste) which suppresses again the lateral bud-growth in much the same way as the terminal bud.

17. Demonstration of effect of kinetin on growth

(*a*) ON CELL DIVISION

EXPERIMENT: 1 to 5 ppm kinetin solution is prepared in distilled water (see Appendix II). Tobacco pith tissue or carrot tissue is grown in 6 petridishes containing White's nutrient medium prepared as follows:

WHITE'S NUTRIENT MEDIUM: Each of the following solutions is separately prepared in a minimum volume of water and then the solutions are mixed one by one in the serial order as arranged below with constant shaking, otherwise precipitation may appear. The final volume is made up to 1 litre. Before use, ten times dilution of the stock solution is necessary.

Serial no.	*Salts*	*Quantity (gm)*
1.	$Ca(NO_3)_2$	2·0
2.	Na_2SO_4	2·0
3.	NaH_2PO_4	0·165
4.	KNO_3	0·80
5.	KCl	0·85
6.	$MnSO_4$	0·45
7.	KI	0·075
8.	$ZnSO_4$	0·150
9.	H_3BO_3	0·150
10.	$MgSO_4$	3·60
11.	$FeCl_3$	a trace

5 ml of 1 ppm, 2 ppm, 3 ppm, 4 ppm, and 5 ppm kinetin solutions are added to the first, second, third, fourth and fifth petridishes respectively. One petridish is kept as control where instead of kinetin solution, 5 ml distilled water is added. Observations are made after seven days.

OBSERVATION: Cell formation is noted in all the treatments (the area of the cell tissue or the per cent of new cells formed may be measured). Maximum cell formation has been observed in the concentration range of 1 to 2 ppm kinetin.

INFERENCE: Kinetin induces cell division. The adenine moiety of kinetin molecule appears essential for this process. The side chain of kinetin may also possess some physical property for regulating the efficiency of it to induce cell division.

N.B. Not only does kinetin promote cell division, but also induce cell enlargement. Kinetin treatment of leaf discs cut from etiolated leaves, e.g., of *Phaseolus* sps. causes significant cell enlargement. The effect of kinetin on cell enlargement is usually associated with IAA and gibberellins.

(*b*) ON LEAF SENESCENCE

EXPERIMENT: *Vigna* or *Phaseolus* seedlings are most suitable for this experiment. 25 ppm kinetin solution is applied on both surfaces of one of the primary leaves leaving the other as such. The trifoliate leaves are removed as soon as they appear. Observation is made from time to time.

OBSERVATION: It is observed that kinetin treated leaf remains green and fresh while the untreated opposite leaf of the pair becomes yellow.

INFERENCE: Kinetin is a potent arrestor of senescence. Leaves treated with kinetin draw metabolites from untreated region and also cause an increase of protein and nucleic acid synthesis. Since kinetin is relatively immobile it is not easily transported to the untreated regions. Hence kinetin treated leaf remains green due to deferal of senescence while the untreated leaf shows the sign of senescence much earlier.

18. Demonstration of effect of gibberellin on growth

(*a*) ON SEEDLING GROWTH

EXPERIMENT: Several suitable seedlings of *Vigna* or *Phaseolus* are grown in two petridishes. One lot of seedlings is sprayed with 20 ppm gibberellic acid (GA_3) solution (see Appendix II) with the help of an atomiser or a sprayer. The other lot is sprayed with distilled water which serves as a control. Observations are made after one week by measuring the height of the seedlings.

OBSERVATION: It is observed that the height of the gibberellin-treated seedlings has increased considerably compared to that of control seedlings.

INFERENCE: Gibberellin, like IAA, promotes cell elongation in the internodal region and thus increases seedling growth. The rate of increase is much higher in case of gibberellin than in IAA.

(*b*) On light-inhibited stem growth

Experiment: Three lots of *Vigna* or *Phaseolus* seedlings are grown in three petridishes. One lot is kept in dark and the other two lots are kept in light of which one lot is sprayed with 20 ppm gibberellic acid (GA_3) solution. The stem growth of the seedlings in each case is measured and compared.

Observation: It is observed that height increases considerably in case of dark-grown and light-grown but gibberellin-treated seedlings compared to only light-grown seedlings.

Inference: Light has an inhibitory effect on stem elongation. Application of gibberellin to the plants growing in light greatly increases their stem growth. Reversal of light induced inhibition of stem elongation by application of gibberellin suggests that endogenous gibberellin is the limiting factor in stem growth. It may be concluded that light causes inhibition of stem growth by lowering the level of endogenous gibberellin and this inhibition is erased by applying exogenous gibberellin to the plants.

19. Inhibition of protoplasmic movement

Experiment: The protoplasm of most living plant cell is frequently in active motion. Sections of *Chara*, *Vallisneria*, *Elodea*, etc. show this movement. Here the granular plastids can be seen to move along this protoplasmic stream (*cyclosis*). Some sections of *Vallisneria* or *Elodea* or *Chara* are put to solutions of respiratory inhibitor like malonic acid, cyanide or uncoupler of ATP like DNP in dilute concentrations. These treated sections are compared with untreated sections.

Observation: It is observed that the rate of movement of plastids or the protoplasmic streaming decreases considerably in case of inhibitor treatment in comparison to controls.

Inference: Though the mechanism of protoplasmic movement is not known yet, the involvement of respiratory energy for this movement is clearly indicated from studies with respiratory inhibitors.

20. Inhibition of water uptake

Experiment: Slices of peeled potato tuber of uniform thickness are prepared. Equal weights of slices are taken in two petridishes, one containing distilled water and the other containing any one of the inhibitor solutions mentioned in the above experiment. The weights of the slices of both the petridishes are taken at an interval of fifteen minutes for a period of two hours. The slices should be blotted carefully before taking weights.

Results: The weight of slices kept in both the petridishes increased gradually but the rate of increase of water uptake is much higher in distilled water medium than that of slices kept in inhibitor solutions.

DISCUSSION: Some amount of water is absorbed by the plant cells by active process, i.e., at the expense of respiratory energy. Treatment of the plant tissue with metabolic inhibitor will affect the process thereby reducing the rate of absorption.

N.B. The active absorption of ions can also be inhibited by using metabolic inhibitors or inhibitors of protein synthesis (chloramphenicol) indicating that this process is dependent upon energy expenditure and enzyme activity. The active accumulation of potassium ions by *Nitella* can be studied by keeping the algae in a dilute solution of KNO_3 as well as by keeping them in the KNO_3 solution containing a metabolic inhibitor. The more potassium ions will be accumulated by the algae, the less will be the conductivity of the external medium which can be detected by a Conductivity meter.

21. Inhibition of transpiration

EXPERIMENT: Two transpiration experiments are set up as in Expt. 8(*c*) of Chapter VI. In one conical flask, instead of water, some inhibitor solution of either respiration (malonic acid) or protein synthesis (chloramphenicol) or uncoupler of oxidative phosphorylation (DNP) or any other antitranspirant like abscisic acid is taken. The rate of transpiration in both the cases is determined.

RESULTS: The rate of transpiration in control set is found to be higher than that of the inhibitor-treated set.

DISCUSSION: Transpiration is dependent upon opening and closing of stomata and the phenomenon of opening and closing of stomata is again an energy-driven enzyme-catalysed process. Thus the rate of transpiration is always inhibited by a respiratory inhibitor which has an effect on oxidative phosphorylation and by inhibitor of protein synthesis which has an effect on enzyme synthesis and activity.

22. Inhibition of respiration

EXPERIMENT: Two experiments on respiration are set up as in Expt. 1 of Chapter X. Two lots of germinating gram seeds of equal weights are taken. One lot of such seed is treated with a dilute solution of any respiratory inhibitor (iodoacetic acid or iodoacetamide or fluoride which are inhibitors of glycolysis; or cyanide, malonic acid, etc. which are inhibitors of Krebs cycle) for 2 hours and the other lot is similarly treated with distilled water only. The rates of respiration in both the lots are measured.

RESULTS: The rate of respiration is much lower in inhibitor-treated seeds than the untreated control.

DISCUSSION: Respiratory inhibitors inhibit enzyme-catalysed reaction as a consequence of which the rate of respiration is inhibited. Iodoacetamide or iodoacetic acid prevents oxidation of triosephosphate to phosphoglyceric acid, fluoride inhibits the enzyme enolase, cyanide inhibits the activity of oxidase enzyme and malonic acid inhibits the enzyme succinic dehydrogenase.

23. Inhibition of photosynthesis

EXPERIMENT: Two experiments on photosynthesis are set up as in Expt. 1(*a*) of Chapter XI. In one beaker, instead of water, weak cyanide solution (inhibitor of dark reaction) or DCMU or CMU solution or atrazine solution (inhibitors of photochemical reaction) is taken. The rate of photosynthesis is measured in both the cases under moderate and uniform light intensity. Other set is kept as control.

RESULTS: The rate of photosynthesis is found to be less in inhibitor treated plants compared to control.

DISCUSSION: Cyanide is an inhibitor of dark reaction, i.e., reductive fixation of CO_2 which is cyanide sensitive. DCMU or CMU inhibits O_2 evolution by chloroplast, i.e., photolysis of water in photosystem II and atrazine inhibits phosphorylation in photosystem I and hence the reduction in the rate of photosynthesis.

24. Inhibition of growth

EXPERIMENT: Some seedlings of *Phaseolus* or *Vigna* are grown in two earthenware pots. 50 ppm solution of any growth retardant (maleic hydrazide, Alar-85, CCC, AMO-1618, etc.) is applied to 10 seedlings of one pot every alternate day for a period of 10 days. The seedlings of the other pot are kept as control. The growth in height of the seedlings as well as leaf areas of the seedlings are measured in treated and untreated controls.

RESULTS: The average height and the leaf area are recorded in treated and untreated controls and results are compared.

DISCUSSION: The growth in height of the seedlings and area of the leaves are much retarded in case of inhibitor-treated seedlings as compared to untreated controls. These growth ratardants mainly act antagonistically with the growth hormones particularly with the gibberellins thereby reducing the growth of the plant as a whole.

25. Inhibition of germination

EXPERIMENT: Some suitable **nondormant and** viable seeds are taken.

Two filter papers are taken in two petridishes. To one petridish 5 ml of any germination inhibitor (coumaric acid or coumarin, abscisic acid, cycloheximide, ferulic acid, etc.) of 10 to 20 ppm concentration is taken. To the other 5 ml of distilled water is added. Twenty seeds are arranged on the filter paper of each petridish and the number of seeds germinated every day is noted for a period of one week.

RESULTS: The speed of germination (number of seeds germinated each day) and the final percentage of germination are recorded in each case.

DISCUSSION: Germination is inhibited by treatment with inhibitors. The inhibitors mainly act on the enzyme system, particularly hydrolases, which takes active role in germination process by breaking down complex substances into simpler ones. These inhibitors may also act antagonistically on gibberellin action which is essential for germination.

26. Inhibition of flowering

EXPERIMENT: Seedlings of *Lactuca* or *Nicotiana* are grown in pots. Some seedlings are treated with 50 ppm of any antigibberellin (AMO-1618, Phosphon-D, CCC, etc.) every alternate day for a period of 10 days. Some seedlings are kept as control.

RESULTS: The days required for bolting (emergence of the inflorescence stalk) is recorded in treated and untreated controls.

DISCUSSION: Gibberellin has an active role in flowering of long day rosette plants. So flowering is inhibited in case of long day rosette plants by application of antigibberellins. However, the mechanism of action of these antigibberellins is not yet clear.

27. Bioassay of natural growth regulators

Assay is a term originally employed for the determination of the proportion of a metal in an ore or an alloy. In recent years it has been applied by biologists to the determination of specific activity of such biological products as hormones, inhibitors, vitamins, antibiotics, etc. The main assay methods which have been employed in determination of growth substances are given below. The details of the methods will be available in any standard text book on growth substances.

A. BIOASSAY OF IAA

(*i*) Coleoptile curvature test by *Avena* coleoptile.

(*ii*) Other curvature tests with cylindrical organs in which cotyledons are removed by an oblique cut from the tops or hypocotyls of *Cephalaria* sp. and ***Raphanus sativa***.

(*iii*) Curvature test with split organs or split pea curvature test with young stem of the garden pea grown in dark for about a week (etiolated).

(*iv*) Segment straight growth test with *Avena* coleoptile.

(*v*) Root growth test with *Triticum vulgare*, *Lens esculantus*, etc.

(*vi*) Epinastic test with tomato leaf.

B. BIOASSAY OF CYTOKININS

(*i*) Callus tissue growth with *Nicotiana tabacum* stem pith, soybean cotyledons, carrot tissue.

(*ii*) Bud formation with protonema of moss.

(*iii*) Leaf expansion with *Phaseolus vulgaris*, radish, etc.

(*iv*) Leaf yellowing test (delay of senescence) with *Xanthium*, barley, oat and bean leaves, radish and cabbage leaf discs.

C. BIOASSAY OF GIBBERELLINS

(*i*) Seedling test with dwarf pea, dwarf maize, seedlings of lettuce, etc.

(*ii*) Leaf growth with segments of immature first leaf base of certain cereals, e.g., *Avena*, *Triticum*, *Hordeum*, etc.

(*iii*) Barley endosperm test with barley endosperm and α-amylase activity.

(*iv*) Leaf senescence test with isolated leaf discs of *Taraxacum* sp., *Rumex* sp. on chlorophyll retention.

D. BIOASSAY OF ABSCISIC ACID

(*i*) Acceleration of abscission in excised abscission zone (explant).

(*ii*) Inhibition of coleoptile curvature or straight growth.

(*iii*) Inhibition of seed germination, including growth of excised embryos.

(*iv*) Inhibition of growth of rice seedling.

E. BIOASSAY OF ETHYLENE

(*i*) Epinasty of leaves. (*ii*) Swelling of stems.

(*iii*) Inhibition of extension growth.

These three responses are together known as 'triple response'. Etiolated pea seedlings are suitable materials for ethylene bioassay.

CHAPTER XIII

GERMINATION, DORMANCY AND FLOWERING

A plant experiences some complicated changes during the repeating cycle of development. This cycle generally starts with germination of seed and continues with the passage of a juvenile phase of growth and the graduation into maturity, followed by a progress into a state of senescence. With maturity the organism is capable of shifting from vegetative to reproductive phase of growth, with initiation and development of flowers, fruits and again the production of a new generation of seeds.

Emergence of radicle from seed coat is called *germination*. In case of seeds in which plumule first emerges, this event should also be considered as germination. During germination of seeds, dormant embryo contained in it starts growing and developing into a new plant which resembles its parent. Some seeds start germinating as soon as water is available, while some other seeds require some periods of suspended growth even when favourable conditions are available. This period is called the *dormancy period* of seeds. Three important events must occur during germination of seeds: (*i*) imbibition and absorption of water, (*ii*) initiation of cell division in the meristematic region of embryo, and (*iii*) enlargement of embryonic cells.

Different kinds of seeds remain viable for different durations of time. Many difficulties are encountered in keeping the seeds in viable condition for maximum length of time, as well as to enable the seeds to germinate when desired. Seeds possessing dormancy may be artificially made to germinate either by physical or chemical treatments.

The flowering process involves a complete change of the products of developing meristems, which instead of giving vegetative parts like leaves, internodes, etc. form the highly specialised structure of flowers and associated appendages. This redirection of growth to the induction of reproductive state in leaves, the initiation of floral meristems, the morphological development of flowers, or anthesis itself, is due to some yet uncharacterised chemical substances of hormonal nature synthesised in the maturing leaves. The triggering of this reproductive developments is usually controlled by different environmental variables such as temperature, light and photoperiods. These are variables which change with seasons, hence permitting a programming of the reproductive activities on a seasonal basis. Photoperiodic conditions are usually perceived by the leaves.

Photoperiodism is a response to the timing of light and darkness. In many species the appropriate photoperiodic schedule need only be given for a few days in order for flowering to occur, even though the plants are subsequently maintained under photoperiods unfavourable for flowering. Such persistent after-effects of photoperiodic treatment are the results of *photoperiodic induction.* The modern concept of the mechanism of photoperiodic induction postulates that two different types of stimulatory factors or hormones take part in flowering process. These two hormones together are called the *florigen complex* which consists of a gibberellin-like hormone and the other an auxin-like, yet unidentified hormone, called *anthesin.* The photostimulation requires the involvement of some pigment present in the leaves. This pigment is called *phytochrome* which can exist in two interconvertible red and far-red light absorbing forms. It is conceived that light or red light (R-form) converts phytochrome to another form (FR-form) and the latter when formed in suitable quantity induces certain dark reactions which result in the production of flowering hormone. The concentration of this hormone formed is dependent on the duration of both light and dark periods. Thus during day far-red form is accumulated in the plant which is inhibitory to flowering in short day plants and stimulatory in long day plants. During dark period far-red form is slowly converted into red-form again which is stimulatory to flowering in short day plants but inhibitory to long day plants.

It is now known that this pigment is responsible for controlling other growth phenomena like germination, dormancy, etc. besides flowering.

According to the photoperiodic requirement plants can be classified into three major classes: (*i*) The *short day plants* (which flower only when the day length is less than a certain critical duration, e.g., *Chrysanthemum, Xanthium, Cosmos, tobacco,* etc.), (*ii*) *The long day plants* (which flower only when the photoperiod is higher than a critical duration, e.g., *Hyoscyamus*, wheat, potato, spinach, etc.) and (*iii*) The *indeterminate* or *day neutral plants* (which have indefinite requirement of photoperiod and flower in all the seasons, e.g., tomato, cotton, pea, etc.).

Like light, temperature also controls flowering in certain species especially in winter varieties. The promotive effect of low temperature on flowering is termed *vernalisation.* The low temperature effect may be obtained in some species when the moistened seeds are chilled or in others when the growing plants are chilled. Some species respond to chilling at either stage. The fact that high temperature treatment could nullify the vernalisation effect establishes the *devernalisation* process, a physiologically significant event. When vernalised grains are exposed to 35°C. for one day the vernalisation effect is erased. It has been postulated that during vernalisation treatment certain chemicals of hormonal nature are produced in the meristematic cells which develop the capacity of differentiation of floral organs. This hypothetical hormone is called *vernalin* by certain authors which is perhaps closely related to gibberellins.

A. Germination

1. Methods of studying germination of seeds

Experiment: There are several suitable methods for studying germination of seeds. Healthy, viable and nondormant seeds of *Oryza*, *Phaseolus*, *Vigna*, cotton, gram and pea are surface sterilised with 1% mercuric chloride for 1 to 2 minutes. The seeds were then washed well with water. The seeds may be allowed to germinate on (*i*) moist filter paper on a petridish, (*ii*) on moist cotton, (*iii*) on moist sand, (*iv*) on commercial vermiculite (germination medium), or (*v*) on net-wire kept on vessel containing water. These are kept in normal laboratory conditions.

Results: Germination behaviour of different types of seeds on different media are observed and recorded. The number of seeds germinated per day may also be noted.

Discussion: The above methods are generally used for germination experiments and in many physiological experiments with seedlings. Some methods are suitable for some experiment while others are not. For simple germination study, moist filter paper or cotton serves best. The method is especially satisfactory when seeds are required in their earlier stages of germination (for enzyme study). For studying seedling growth moist sand or vermiculite media are suitable. To study the root growth of the seedling the net-wire may be conveniently used. The seedling can be removed easily from such a germinator by carefully cutting the meshes of net. It should be remembered that all the media be kept moist at all times and that these be kept within a favourable range of temperatures.

N.B. (*i*) The rate of germination, if slow, can be hastened by artificial, physical and chemical treatments (variable for different seeds) like red light, dark, temperature (high or low), KNO_3, thiourea, gibberellin, etc.

(*ii*) *Germination capacity* or *germinability* of seeds can be calculated from the above experiment. It is the percentage of seeds capable of germination under favourable conditions. It is determined at a certain specified time after sowing, called the final or ultimate germination.

(*iii*) *Germination speed, rate* or *energy:* This is calculated from the time taken by the seeds to germinate and can be expressed in the following ways:

(*a*) As the proportion of seeds germinated by certain time after sowing.

(*b*) As the time required to reach 50% of germination capacity or of final germination.

(*c*) As a special figure which takes into account the time taken by each seed to germinate, e.g., *coefficient of velocity* (Kozlowski, 1926) which expresses the mean germination of a sample by integrating the germination times of all individual seeds:

$$\text{Coefficient of velocity (CV)} = \frac{\Sigma n \times 100}{\Sigma(n \times Dn)}$$

where n = number of seeds germinated on day Dn,

Dn = number of days after sowing (day of sowing = D_o).

2. Methods of testing viability of seeds

It is often of great importance to know the viability of different seed samples. The percentage of viability is usually determined directly by germination tests with small samples. Many kinds of seeds are perfectly viable but may be very slow in germination or may be dormant. For such seeds an indirect and more rapid method of testing viability would be desirable. Many such methods have been used with varying degrees of success. Some methods are summarised below.

(*a*) Respiration test: Respiration is an indication of life but such tests are complicated by the fact that dormant or resting seeds show very low respiration and micro-organisms growing on seeds may show respiration when seeds themselves are all dead.

(*b*) Electrical conductance method: Some authors have observed that if seeds are soaked in distilled water and the electrical conductance of the bathing solution is then tested, the increase in conductance is roughly proportional to the percentage of dead seeds. This increase in conductivity is due to leaching of metabolites from dead seeds which become pervious owing to increasing permeability.

(*c*) Potassium permanganate method: Using weak solution of $KMnO_4$ it has been observed that the decolourisation of this solution is increased as the proportion of the dead seeds in the sample increases. It is a well-known fact that dead cells become freely permeable to their contained solutes which leach out into the bathing solution, whereas leaching from living cells is very much less. This increased permeability on death undoubtedly accounts for the increased content of electrolytes (as in *b*) and reducing substances in the bathing solution resulting in decolourisation of $KMnO_4$ solution.

(*d*) Indigocarmine method: A still simpler method of testing of increased permeability of the dead seeds involves soaking for a few hours in certain aniline dyes, such as indigocarmine which will stain dead cells but not living ones. For applying this and similar tests it may be necessary to remove nonliving seed coats which show low permeability.

(*e*) EMBRYO CULTURE METHOD: A method that has proved successful in testing the viability of seeds having prolonged dormancy involves removal of the embryo from its cotyledon or endosperm and placing the naked embryo on granulated peat moss or on sterilised nutrient agar medium (White's nutrient medium). By this method viability may be tested within seven to ten days. The nonviable seeds fail to germinate.

(*f*) TETRAZOLIUM TEST: A common and generally adopted method of testing viability of seeds involves soaking of seeds in 1 to 2% tetrazolium chloride solution (2, 3, 5-triphenyltetrazolium chloride). The living and viable seeds take bright red colouration (more intense in embryo) while the nonviable seeds do not. Tetrazolium chloride solution is colourless but forms carmine red *formazan* upon reduction. The salt is thus an oxidation-reduction indicator and the development of the nondiffusible red colour in a specific tissue is presumably an indication of the presence of active respiratory processes in which hydrogen ions (due to dehydrogenase activity in respiration) are transferred to the tetrazolium chloride causing its reduction. This process is of general value in physiological experiments as well as an index of germinability of seeds.

3. Effect of presoaking in water on germination of seeds

EXPERIMENT: Perfectly viable seeds of (*i*) pea, (*ii*) bean and (*iii*) rice (125 each) are soaked in distilled water in three petridishes. A lot of 25 seeds from each petridish is taken out at intervals of 0, 8, 24, 48 and 72 hours and allowed to germinate in dry petridishes. The number of seeds germinated after three days in each case is recorded.

RESULTS: The number of seeds germinated after different periods of presoaking in water is tabulated in each case and percentage is calculated.

DISCUSSION: Absorption of water being the first prerequisite of germination, it is expected that presoaking of seeds in distilled water should accelerate germination. Within limit, this is correct but the effects depend to a great extent on duration of presoaking, soaking condition, size of seed and other properties of seeds. Excessive presoaking frequently results in decrease of germination percentage due to anaerobic condition of the cellular environment.

4. Effect of temperature on germination of seeds

EXPERIMENT: Perfectly viable healthy seeds of rice, wheat and cucumber are taken for this experiment. Twenty-five seeds of each type are sown in well moistened sand contained in three petridishes. Petridishes are then placed at a temperature of 10°C. ± 2 (refrigerator), 25°C. ± 2 (at room) and 40°C. ± 2 (incubator). Petridishes are kept under dark condition. The percentage of germination is noted every alternate day for a period of 10 days.

RESULTS: The rate of germination for each type of seed at a particular temperature is graphically plotted.

DISCUSSION: The rate of germination for each type of seed is expected to be maximum at 40°C., optimum at 25°C. and minimum at 10°C. In the range of temperature within which certain seeds germinate there is usually an optimal temperature below and above which germination is delayed but not prevented. The optimal temperature may be taken to be that at which the highest percentage of germination is attained in the shortest time. There may be three temperatures where germination may be found to be minimum, optimum and maximum (Sachs 1860, introduced the concept of three cardinal points for these temperatures).

5. Effect of light on germination of seeds

EXPERIMENT: Perfectly viable healthy seeds of *Ruellia tuberosa*, *Mirabilis jalapa* and *Datura stramonium* are taken for this experiment. Twenty-five seeds of each are soaked in water for an hour and placed on moistened filter paper in petridishes. Each type of seeds is placed in three petridishes and is kept in continuous light and continuous dark. The third serves as control (12 hours light and 12 hours dark). The rate of germination is observed from time to time for a period of one week.

RESULTS: Rate of germination of seeds under light, dark and control conditions is calculated and graphically plotted.

DISCUSSION: Light influences germination of many seeds. Seeds may be divided into three categories, viz., (*i*) *negatively photoblastic* seeds which germinate only in dark, (*ii*) *positively photoblastic* seeds which germinate only in light, and (*iii*) *photoinsensitive* seeds which are indifferent to the presence or absence of light in their germination responses. In the present experiment *Mirabilis jalapa* falls to the category (*i*) because its germination is favoured by dark only, *Ruellia tuberosa* falls to the category (*ii*) because its germination is favoured by light only, while the germination of *Datura stramonium* is indifferent to either light or dark and belongs to the category (*iii*).

6. Effect of corbon dioxide on germination of seeds

EXPERIMENT: Using the same type of apparatus and procedure employed in Expt. 11 of Chapter X, the effect of CO_2 concentration on germination of suitable seeds is observed. The rice seeds are first soaked in water before starting the experiment and then placed in petridishes under bell jars containing different concentration of CO_2 (different concentrations are made by allowing CO_2 to enter the bell jar for different durations).

One petridish containing such seeds is placed under bell jar containing normal CO_2 which serves as control. The percentage of germination is observed every alternate day for a period of seven days.

RESULTS: The rate of germination in different concentrations of CO_2 is noted and graphically plotted.

DISCUSSION: Most seeds fail to germinate if the CO_2 tension is greatly increased. The response of germination of seeds towards CO_2 concentration is variable from species to species. Concentration of CO_2 in excess of 10% is in most cases inhibitory to germination. Inhibition of germination of seeds is due to the toxic effect of CO_2 on respiration.

N.B. The effect of oxygen concentration may also be studied provided arrangement for O_2 supply is available. The effect of O_2 is usually promotive within a particular limit.

B. DORMANCY

7. To study the phenomenon of dormancy

(I) SEED DORMANCY

EXPERIMENT: Freshly harvested seeds of *Xanthium*, *Nicotiana* and *Aman* variety of rice (e.g., Nagra, Kalma, Kumargore, etc.) are sun-dried for 2–3 days to bring down the moisture level to 11–12 percent. Fifty seeds of each are surface sterilised with mercuric chloride (1% for 1 minute) and allowed to germinate in petridishes containing moistened filter paper. Petridishes are kept at laboratory temperature of 25–30°C. After seven days the percentage of germination of seeds is recorded in each case.

RESULTS: Number of seeds germinated in each case is determined and percentage of seeds germinated is calculated.

DISCUSSION: Many seeds do not germinate when placed under conditions which are normally favourable for germination (adequate water supply, suitable temperature, normal atmospheric composition). These seeds are viable as they can be induced to germinate by various treatments. Such seeds are said to be dormant. Dormancy is due to various causes such as immaturity of embryo, impermeability of seed coat to water or gases, prevention of embryo development to mechanical causes, special requirements for temperature or light or presence of substances inhibiting germination. In the present experiment the seeds are found to be very poor in germination (i.e., less than 20 per cent or so) though the external conditions have been favourable. Hence the seeds may be called dormant.

(II) BUD DORMANCY: Refer Expt. 15 of Chapter XII.

8. Methods of breaking seed dormancy

The dormancy of seeds can be erased artificially by various special treatments which may be grouped into two broad categories: (*a*) *Physical treatment*, and (*b*) *Chemical treatment*. Some treatments are found to be most suitable for some species for breaking dormancy while others are not. The effective dormancy breaking agent is to be found out by trials on seeds of a species having dormancy. Most widely used treatments are enlisted below:

PHYSICAL TREATMENTS

(*i*) Heat treatment at 40°C. to 45°C. for different durations.

(*ii*) Low temperature treatment at 2°C. to 8°C. (seeds presoaked for 36 hours) for 12 to 24 hours.

(*iii*) Alternate heating and cooling for several times.

(*iv*) Alternate drying and wetting for several times.

(*v*) Soaked seeds (24 hours) exposed to red light for 1 to 2 hours at 15–25°C.

(*vi*) By dehusking of seeds or removal of seed coat.

CHEMICAL TREATMENTS

INORGANIC CHEMICALS

(*i*) By acid treatment: Dilute solutions of HNO_3, HCl or H_2SO_4 (0·1–0·5%) for different durations in minute.

(*ii*) By KNO_3 (1–3%), NH_4NO_3 (1–3%), H_2O_2, H_3BO_4, etc.

(*iii*) By gasses: By increasing oxygen concentration.

ORGANIC CHEMICALS

(*i*) NON-HORMONAL: Thiourea, KSCN, Ascorbic acid, etc. (10 to 100 ppm).

(*ii*) HORMONAL: Gibberellic acid (1 to 100 ppm), Kinetin (1 to 100 ppm), Ethylene (Ethrel solution of 100 to 300 ppm).

EXPERIMENT: The efficiency of the above methods for breaking dormancy of seeds referred in Expt. 7 may be determined by trials.

DISCUSSION: There are three general types of internal control of dormancy. (*i*) Limitations of permeability, (*ii*) limitations by growth substances including inhibitors and their counteracting of promotive substances and the activation or inactivation of promoters and inhibitors by temperature or light, and (*iii*) physical limitations on the enlargement of the embryo and its emergence. It appears that most dormancy breaking treatments can be roughly accounted for on the basis of these three types of controls. For example, scarification treatment (physical treatment) alters the first and the third types of limitations; low temperature and high oxygen treatments facilitate the water and gas entry and affect the inhibitor content which may be present. Inhibitor accumulation may be a consequence of high temperature experience. Gibberellins, cytokinins and nitrates may also modify the effective inhibitor levels in various ways and other chemicals which break dormancy may act both on the effective inhibitor level or on water or gas entry step through physical alterations of seed coat and metabolic alterations of the tissues concerned.

C. Flowering

9. Demonstration of photoperiodic response of plants towards flowering

Experiment: The following short-day, long-day and day-neutral plants are selected.

Short-day plants (SDP): *Chrysanthemum* or *Xanthium* or *Aman* variety of rice or *Cosmos* or *Nicotiana tabacum.*

Long-day plants (LDP): Spinach or wheat or potato or radish or winter barley or *Hyoscyamus niger*, or *Nicotiana plumbaginifolia.*

Day-neutral plants (DNP): Tomato or maize or cucumber.

The plants are grown in pots and when the seedlings are grown to a certain height these are given photoperiodic treatment, i.e., half of the plants consisting of all the three types is exposed to long-day (16 hours light and 8 hours dark), and the other half consisting of the same types is exposed to short-day (16 hours dark and 8 hours light). Long-day treatment may be given by keeping the plants at a distance of 3 feet from a 200 W lamp. The position of the pots is to be changed frequently so that the plants get uniform light. The dark treatments may be given by keeping the pots in dark room or keeping them under black tent in the field. The experiment is continued till flowering is achieved in at least some of the plants and photoperiodic type of the plants is determined from such treatments.

Results: It is observed that the long-day plants flower only where long-day exposure is given but remain in vegetative state in short-day treatment. The reverse result is obtained in case of short-day plants, i.e., plants flower only in short-day condition but not in long-day treatment. In case of day-neutral plants it is observed that they flower in both the conditions.

Discussion: Short-day plants are those which flower only when exposed to day lengths of less than a certain critical duration (generally less than 12 hours). The long-day plants are those which flower only when exposed to photoperiods greater than a critical duration (generally greater than 12 hours). It has been found that the flowering is not only dependent on definite light period but also on definite dark period. Consequently, the short-day plants may be considered as long-night plants. The dark period requirement is more important so far the delimitations of plants on the basis of flowering response is concerned.

10. Demonstration of photoperiodic induction of plants towards flowering

EXPERIMENT: Several short-day and several long-day plants are selected and grown as in Expt. 9. A few short-day plants of uniform age of each type are given one photo-inductive cycle (16 hours dark + 8 hours light) and then transferred to long-days. Again a few long-day plants are given one photo-inductive cycle (16 hours light + 8 hours dark) and then transferred to short-days. Now the number of photo-inductive cycles is increased by one day up to ten days in case of other plants and are similarly transferred to their opposite photoperiodic conditions. The flowering is observed in each case. If the plants do not flower even after giving 10 photo-inductive cycles the number of cycles may be increased.

RESULTS: From the experiment the minimum number of photo-inductive cycles required by each type of plant for flowering is determined.

DISCUSSION: The relative length and sequence of both light period and dark period influence the flowering of plants. There are two stages in flowering, first the differentiation of floral primordia in the meristematic apex and then the development of mature flowers. Different stages of the reproductive growth may have their own specific requirements of photo-inductive cycles, i.e., light and dark periods in alternating sequence. Long-day plants naturally flower during spring and summer while the short-day plants flower during winter and autumn. Plants can be made to flower by artificially giving any desired cycle of light and dark periods. The minimum requirement of this cycle varies from species to species. This is called *photoperiodic induction.*

11. To demonstrate that leaf is the locus of stimulus of flowering

EXPERIMENT: Suitable short-day plants of uniform age are selected for this experiment. One lot of four plants is completely defoliated. Leaves of another lot of four plants are defoliated leaving only one fully mature leaf at the lower node of each plant. Other lot of four plants is kept as control. All the plants are now given appropriate short-days till flowering. (This experiment can be repeated with any suitable long-day plant.)

RESULTS: Flowering is noted in all the plants excepting the lot where all the leaves have been defoliated or removed.

DISCUSSION: For both long-day and short-day plants it is known that the stimulus of photoperiod is received by the leaves and not by the growing points where the flower appears. Plant whose leaves have all been removed fail to produce flowers even when given the appropriate conditions which induce flowering in intact plants or plants having at least one leaf. It is now known that the interconvertible pigment phytochrome is involved in initiating floral stimulus which is mediated by some chemical substances of hormonal nature and this pigment is present in the leaves.

12. Effect of red and far-red light on flowering

EXPERIMENT: Short-day *Xanthium* plant is conveniently used for the experiment. Arrangements are made for red and far-red treatments of the plants in the following way. A white bulb (100 W) is covered with double layered red celophane paper which acts as a red light source and a white bulb covered with blue and red celophane papers serves as a far-red source. A set of four plants is kept under red light, a second set under far-red light, a third set first under red light for 6 hours and then transferred to far-red light, and the fourth set is kept under far-red light for 6 hours (critical photoperiod for *Xanthium* is 8·5 hours) and then transferred to red light.

RESULTS: Plants under continuous red light do not flower but under continuous far-red light flower. Plant, treated with red light followed by far-red also flower while plants treated with far-red light followed by red light do not.

DISCUSSION: Study of the action spectrum of light flash given during dark period to inhibit flowering in *Xanthium* reveals that red light (640 to 680 nm, nm=mμ) is most effective in causing inhibition. It is also found that the effect of red light can be reversed or nullified by far-red light (710 to 740 nm). If far-red flash is again followed by red flash flowering is again inhibited. The quality of the light experienced last by the plants determines the flowering response. The fact that the red light effects on plants are so often reversible with far-red light and that the extracted pigment (phytochrome) shows a spectral shift after exposure to red or far-red light, establishes that there are two forms of photochrome—a *red absorbing form* and a *far-red absorbing form.* Exposure to red light preferentially activates the red absorbing form and converts it into the far-red absorbing form. Exposure to far-red light has the reverse effect. The manner in which the two forms of this pigment regulate the physiological process of flowering in plants is not fully known.

13. Effect of low temperature (vernalisation) on flowering

EXPERIMENT: Some wheat seeds are sown in pots during autumn (October and November) and some seeds are sown during spring (March and April). A few seeds are allowed to germinate partially and these are then kept at cold temperatnre (4 to 5°C) for a period of 12 to 14 weeks during January to February. After this the partially germinated seeds are sown into pots during spring at normal temperature.

RESULTS: The plants raised from seeds sown during autumn and raised from the low temperature-treated seeds sown in spring flower during the following summer whereas the plants grown from the unvernalised seeds sown in spring do not flower in the following summer.

DISCUSSION: Many species are induced or promoted to flower by low temperature especially many biennial and perinnial plants. Less common are species which are caused to flower by high temperatures. The promotive effects of low temperature on flowering is termed vernalisation. By vernalisation treatment of plants or by normal fulfilment of the cold requirement of the plant, certain chemicals of hormonal nature are produced in the meristematic cells which develop the capacity of differentiation of floral organs. Some authors have postulated that a hormone called *vernalin* is produced in meristems of shoot apex and embryos by vernalisation treatment. When this hormone is synthesised in suitable amounts, the shoot apex differentiates into floral organs instead of forming vegetative shoots.

14. Effect of gibberellin on flowering of long-day plants

EXPERIMENT: Moderately matured plants of *Lactuca sativa* and *Nicotiana plumbaginifolia* are selected for this experiment. 250 ml of 100 ppm aqueous solution of gibberellic acid (GA_3) is prepared. The plants are either sprayed with GA_3 solution with the help of holm sprayer or atomiser or GA_3 may be applied to the leaves with the help of a brush for 7 days consecutively. A few plants of each are maintained as control. All the plants are kept at normal environment.

RESULTS: The time of flowering of all the treated and control plants is recorded.

DISCUSSION: Application of GA_3 causes rosette plants to emerge out the inflorescence stalk (bolting) and then to flower. The stimulation of flowering with GA_3 can be asigned to two groups of plants; those that are caused to flower by low temperature and those which flower under long-day condition. In each case where GA_3 is effective, the species is rosette type which generally bolts before flowering; species which are long-day plants and are not rosette generally do not respond to GA_3.

N.B. Flowering may be controlled by regulating C/N ratio as follows. Very low carbohydrate, high nitrogen—no flowering and no fruiting; moderate carbohydrate, high nitrogen—no flowering, no fruiting; high carbohydrate and moderate nitrogen—abundant flowering and high fruiting; very high carbohydrate and low nitrogen—little or no flowering and fruiting.

The flowering of a few plants can be rather precisely initiated by treatments with ethylene, acetylene, NAA or 2, 4-D and with similar compounds.

CHAPTER XIV

PLANT MOVEMENT

By movement it is meant the change in position of an organism or a part of its body. In a strict sense of the term, the movement is not usually associated with plants particularly with trees or shrubs. If movement is thought in the narrow sense of an entire organism moving, then very few instances are found in plant world. Perhaps the best known examples are the unicellular alga *Chlamydomonus* which moves towards light or when spermatozoid moves towards the egg cell by means of flagella. Again if the movement is thought in the sense of a part of an organism moving due to unequal growth rates of the opposite side, then numerous examples can be cited, viz., parts of the plants, their roots, their leaves, their stems all are capable of movement.

Plant movement may be classified into two broad divisions on the basis of convenience rather than on principle. A. *Autonomic* or *spontaneous movement*—this is purely a mechanical movement unrelated to any irritability of the protoplasm. Movements associated with dehiscence of sporangia and with explosive bursting of many fruits, dispersal of spores and seeds, movement of elaters are some of the examples. B. *Induced or paratonic movement*—these movements are induced when stimuli act on irritable or sensitive protoplasm. Induced movement is classified as follows:

(*i*) **Movements of locomotion**—Movement shown by freely motile organism and free parts of fixed plants.

(*a*) Movement by internal stimulus: Examples are movements of some motile algae, fungi, spores in living cells of higher plants; the movements of chromosome during nuclear divisions.

(*b*) Movement by external stimulus: Examples of *Phototaxis* of phototrophic bacteria and motile algae; *Chemotaxis* of spermatozoids towards the ovum; *Thermotaxis*, etc.

(*ii*) **Movements of orientation or curvature:** Movement shown by attached organ of fixed plants.

(*a*) Movement by internal stimulus: Examples are growth movements, e.g., growth itself, nutations; epinasty and hyponasty and movements of variations, e.g., pulvinar movement of lateral leaflets of *Desmodium gyrans*.

(*b*) Movement by external stimulus: This is again divided into two groups.

(*i*) Tropic movement: Direction of response bears a definite relation to the direction of the stimulus, generally either towards it or away from it. Stimulus acting on the protoplasm from one direction only. Examples are *Phototropism*, *Geotropism*, *Hydrotropism*, *Chemotropism*, etc.

(*ii*) NASTIC MOVEMENTS: Here the response is unrelated to the direction of the stimulus. In this case direction of movement always depends on the reacting organs and not only on environment. Stimulus acting on the protoplasm from all sides. Examples are *Photonasty*, *Thermonasty*, *Seismonasty*, *Nyctinasty*, etc.

The environmental factors responsible for growth difference are known as stimulus. The length of time of which the plant or plant organ must be continuously exposed to a given intensity of stimulus for visible response to follow is called the *presentation time;* the period from the commencement of stimulation to the beginning of reaction or response is termed *reaction time* and the time it takes for the plant to recover its original position after the stimulus has been removed is the *relaxation time.*

1. Demonstration of autonomic movement

(*a*) BURSTING OF FERN SPORANGIUM

EXPERIMENT: Some mature brown and some immature fern sporangia from sori are mounted on a slide with a drop of water and observed under the microscope after carefully covering them with cover glass.

OBSERVATION: It is observed that mature sporangia dehisces at the region of annulus and spores are liberated. Whereas immature sporangia do not burst at all.

INFERENCE: The autonomus movement of the sporangium causes bursting of it for dispersal of spores.

(*b*) BURSTING OF FRUITS

EXPERIMENT: Some mature (black) fruits of *Ruelia tuberosa* are collected and put in a petridish containing water.

OBSERVATION: It is observed that the fruits burst open to liberate seeds.

INFERENCE: The autonomus movement of fruit caused dispersal of seeds. It is a purely mechanical movement unrelated to any irritability of protoplasm, rather this is purely a hygroscopic movement largely determined by presence or absence of water.

(*c*) MOVEMENT OF OSCILLATORIA

EXPERIMENT: Some freshly collected *Oscillatoria* filaments are mounted on a slide and observed under the microscope.

Observation: It is observed that the ends of the filament of the alga move to and fro.

Inference: Due to the autonomous movement the filament moves.

2. Demonstration of induced movement

A. Movement of locomotion

(a) Movement by internal stimulus

Experiment: A thin section of *Vallisneria* leaf is mounted on a slide and observed under the microscope.

Observation: Protoplasmic streaming movement (shown by the movement of cytoplasmic granular plastids which are actually taking along the stream) is observed.

Inference: This movement is due to internal stimulus. But the actual mechanism of this movement is not known. However, expenditure of respiratory energy is thought to be involved.

(b) Movement by external stimulus

Experiment: A fern prothallus is washed and slightly dried and mounted on a slide in a drop of water under coverglass. Now a drop of sodium malate solution (1%) is put on the edge of the cover glass and allowed to come in contact with the prothallus.

Observation: When observed under the microscope it is found that the antherozoids show movement being induced by sodium malate.

Inference: This type of movement is induced by a substance which is chemical in nature. So it is called chemotactic movement.

B. Movement of curvature

(a) Movement by internal stimulus

(*i*) **Experiment**: This movement is exhibited by leaflets of telegraph plant, *Desmodium gyrans* (Figure 39) or by the leaflets of *Oxalis*.

Observation: It is observed that two small lateral leaflets of *Desmodium* or *Oxalis* display periodic movement.

Inference: The movement shown by the two small lateral *Desmodium* leaflets is effected by periodic

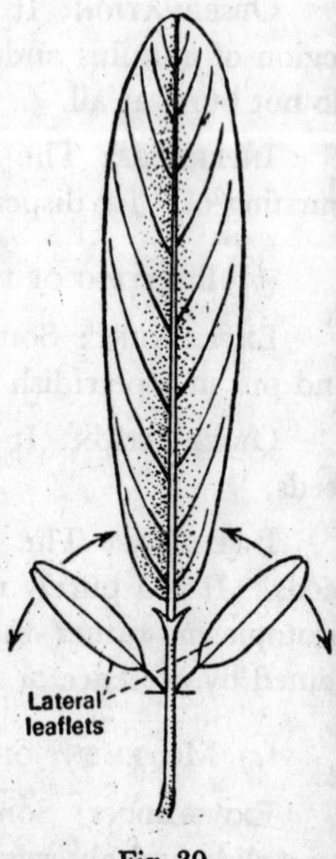

Fig. 39.

changes in turgidity of the pulvinus of the leaf stalk. The movement in the leaflets of *Oxalis* is also due to periodic changes in turgidity of cells of pulvinus.

(*ii*) EXPERIMENT: The best example of nutation movement is exhibited by twining plant like *Cuscuta*, *Pisum*, etc. This movement is also exhibited by stem tips in course of their elongation, showing an irregular spiral path. The following experiment is performed to demonstrate this movement.

A narrow glass thread is prepared from a glass rod and about 5 cm portion of this thread is attached vertically to the stem tip of a vigorously growing potted plant by means of sealing wax. A white square paper (2 cm × 2 cm) is perforated in the middle and passed through the glass thread and kept in a position near the base of the thread. Now a rectangular glass plate is horizontally placed on a ring stand and its height is so adjusted that the tip of the glass thread just remains below the glass plate. The position of the tip of the glass thread is observed from time to time from top of the glass plate and its position is marked on the plate with India Ink (Figure 40).

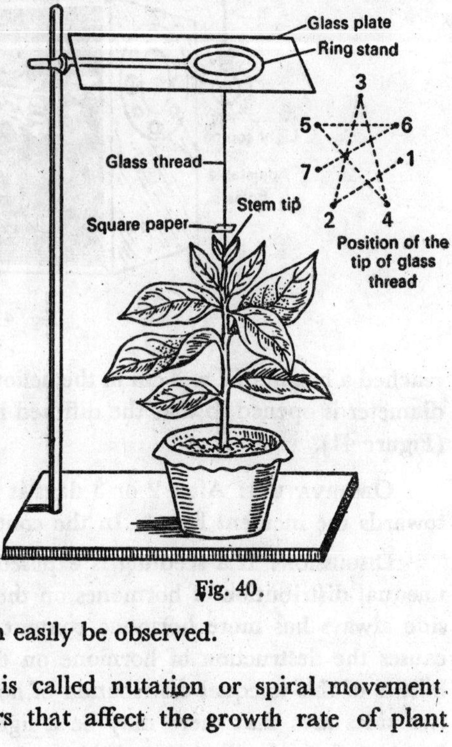

Fig. 40.

OBSERVATION: By plotting the dots at different time intervals the direction of movement can easily be observed.

INFERENCE: This movement is called nutation or spiral movement. It is due to some internal factors that affect the growth rate of plant segments.

N.B. The twinning of tendrils along the support may be explained on the basis of the fact that the tissues of the tendril which are touching the support grow at a faster rate than those which are away from the support. Because of this uneven growth on two sides of the tendrils, a curvature results and the tendrils twine around the support. This movement is under the control of hormones and stress-ethylene production.

(b) MOVEMENT BY EXTERNAL STIMULUS

TROPIC MOVEMENTS

(i) PHOTOTROPISM

EXPERIMENT: Some *Phaseolus* seedlings are grown in a pot which placed at the bottom of a heliotropic chamber (a wooden cabinet havin holes of different diameters which can be closed or opened as desired). A first the holes of the chamber are closed to cut off light source. Som control plants may be grown in a dark room. When the seedlings hav

Fig. 41.

reached a height of 2 to 3 cm in the heliotropic chamber, one hole of suitabl diameter is opened so that the diffused light comes from one direction onl (Figure 41).

OBSERVATION: After 2 or 3 days it is observed that the seedlings ben towards the incident light. In the control, the seedlings grow straight.

DISCUSSION: If a seedling is exposed to unilateral light it results in a unequal distribution of hormones on the two sides of the tip. The darke side always has more hormone content than the illuminated side. Ligh causes the destruction of hormone on the illuminated side. Other possi bilities of this unequal distribution of hormone in different amount on th two sides are: that there may be a light induced transverse migration c hormone from the illuminated side to dark side or that the sensitivity of th tissue to hormone on the lighted side may be decreased. The hormon diffuses downwards and reaches the zone of response in unequal amount causing different rates of cell elongation which in turn causes curvature t take place.

N.B. Effect of different amounts of light may be studied by adjustin the holes of different diameters.

(*ii*) GEOTROPISM

(*a*) EXPERIMENT: Some gram seeds are allowed to germinate until the plumules are 1 to 2 cm long. The inner side of a glass jar is covered with moist cotton. The seedlings are arranged in different positions on the moist cotton so that some have the plumular end downward, others have the plumular end upward while the others have the plumular end in horizontal position.

OBSERVATION: It is observed that in all the cases the radicle grows downwards and plumule upwards.

INFERENCE: The primary root is positively geotropic, i.e., bends towards the centre of the earth and stem is negatively geotropic, i.e., grows vertically upwards. The stimulus of gravity must be unilateral since the force is exerted from one direction only. When such organs are forced out of their upright position, they may assume it again if they are still capable of growth. The geotropic movement is due to unequal growth which is caused by an unequal distribution of hormones. The reaction of roots to hormones differs from that of stem. In the stem the site containing more hormone grows fast, the opposite is the case in the root. A particular concentration of hormone induces acceleration of growth in the shoot end while the same concentration may produce an inhibitory effect in the growth and enlargement of the root cells. The hormone in this case concentrates on the lower side of the plant and possibly growth inhibitors accumulate on the upper part of the horizontally placed mesocotyl resulting in more rapid growth of the cells of the lower side of the stem tissue causing them to bend upwards (negatively geotropic) and faster growth on the upper side of the root tissue, causing roots to bend downwards.

(*b*) EXPERIMENT: When vertically erect shoots of many plants are kept horizontally, their shoot apices bend upward to grow vertically. That the force of gravity acts as a stimulus in bringing about geotropic curvature can be proved by means of a *clinostat* (Figure 42). This instrument is of

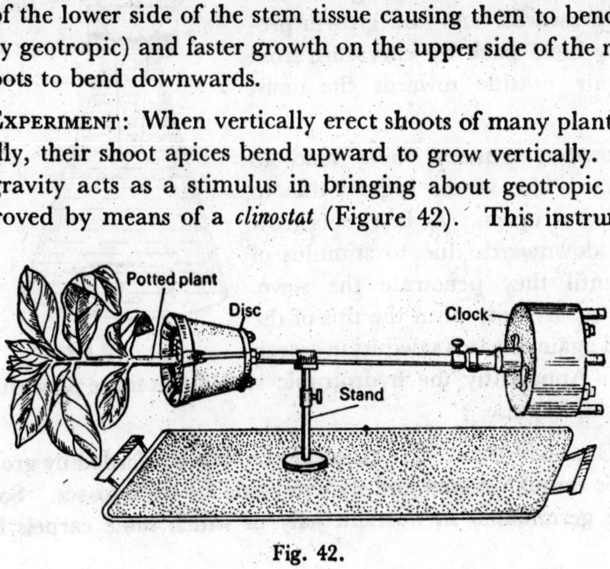

Fig. 42.

many types, but the main thing is that they have a clock to the axis of which is attached a disc (perpendicularly attached at the centre) that can be rotated horizontally either slowly or rapidly with the clock. A pot containing a seedling is fixed on the disc and the whole apparatus is placed in a horizontal position and the clock is started so that the pot rotates one to four revolutions per hour.

OBSERVATION: After two or three days it is observed that the shoots and roots keep on growing in horizontal directions without bending upward or downward.

DISCUSSION: This is because the gravitational stimulus is successively perceived equally on all the sides of the stem or the root as they rotate, keeping all the sides to grow at the same rate and thus the organ grows in horizontal direction. In this condition the gravitational stimulus does not fall unilaterally on the same side of the organ. However, if the clock is stopped the gravitational stimulus becomes unilateral and causes unequal growth so that the shoot moves upward and the root downwards.

(*iii*) HYDROTROPISM

EXPERIMENT: The positive hydrotropism of roots may be demonstrated by growing gram seeds in moist saw dust in a shallow perforated funnel placed vertically on a conical flask (Figure 43).

OBSERVATION: The roots grow vertically downwards. As the growth proceeds they bend back by curvature from the dry air outside towards the moist saw dust.

DISCUSSION: Primary roots tend to grow towards the wetter regions due to positive hydrotropism. At first roots grow vertically downwards due to stimulus of gravity until they penetrate the sieve. But due to hydrotropism the tips of the roots bend again towards sawdust in search of water. Apparently the hydrotropic stimulus is more powerful than the gravitational impulse.

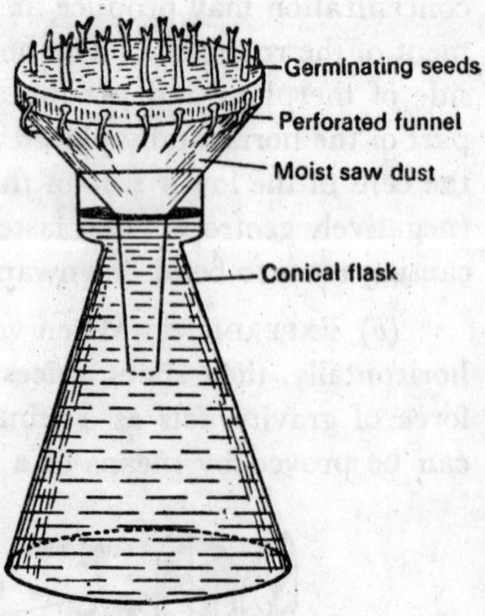

Fig. 43.

(*iv*) CHEMOTROPISM: The classical example is afforded by growth of the pollen tube through stigma and style towards the embryosacs. Some pollen grains are germinated in nutrient jelly in which some carpels have been sown.

OBSERVATION: Germinating pollen tubes grow away from the air and move towards the pieces of carpel.

DISCUSSION: The pollen tubes show positive chemotropism. The first orientation of germinated pollen tube into the stigma takes place by hydrotropism, followed by a mechanical conduction all along the slimy way. For the final part of the movement of pollen tube through the style the chemotropic orientation is responsible.

NASTIC MOVEMENTS

(v) PHOTONASTY: Opening of some flowers during day time is due to photonasty. For example, flowers of *Pentaptes phoenicia* fully open during noon but remain closed during night. The opening and closing of leaves of many leguminous plants may be studied. The opening of stomata in light and closing in dark is also due to photonastic movement. In photonastic movement light from all directions acts as a stimulus which causes opening and closing of flower petals, leaflets and stomata. Light and darkness cause variation in turgidity of the cells—hence the movement.

(vi) NYCTINASTY: The opening and closing of certain flowers, e.g., *Cactus*, *Cestrum*, *Nyctanthes*, *Mirabilis*, etc., are due to nyctinasty. Here changes in temperature and light during the day or night induce visible response which may be termed as nyctinasty. Changes in turgour pressure are also involved here.

(vii) SEISMONASTY: This phenomenon is observed in the movements of leaflets of *Mimosa pudica* in response to touch, shock, etc. When a leaf of *M. pudica* is suddenly touched the leaf and its parts droop down rapidly due to shock stimulus. If the shock is strong the entire leaf droops down. The whole thing consists of three movements—movement of main petiole, movement of secondary petiole and closing and folding up of respective leaflets. The bases of the main and secondary petiole are provided with pulvini consisting of loose parenchymatous cells with intercellular spaces inside and the movements are brought about by changes in the turgour of the cells of pulvinus.

CHAPTER XV

PLANT REPRODUCTION

Since the life of an individual plant is limited in duration it has developed various mechanisms to continue the perpetuation of the species and also to multiply its number. It is certainly an essential criterion of life. The main feature of reproduction consists in a portion of an individual continuing after its death and possessing the power of developing into new individuals. The principal methods by which an individual reproduces itself are *vegetative*, *asexual* and *sexual*.

In vegetative reproduction there are two main methods of propagation: natural and artificial. In natural methods a portion gets detached from the body of the mother plant and this detached portion embarks upon a new career under suitable conditions. Gradually it grows up to a new independent plant. The methods by which the vegetative propagation takes place are many and varied. In artificial methods a portion can be separated only by special technique from the body of the mother plant and grown independently. There are several such methods.

The asexual reproduction takes place by means of special cells or asexual reproductive units cut off by the parent plant to grow by themselves into new individuals without two such cells fusing together as in sexual reproduction. Asexual reproduction takes place by two methods: (*i*) by fission and (*ii*) by spore formation.

The sexual reproduction consists in the fusion of two sexual reproductive units called gametes. Gametes are always unicellular and microscopic in size. They cannot grow by themselves but acquire the power of growth only after two gametes of opposite sexes have fused together. The product of fusion is called zygote which develops into a new plant. The reproduction in which pairing gametes are similar is known as *conjugation* and in which the pairing gametes are dissimilar is known as *fertilisation*.

1. Vegetative reproduction

A. Natural methods of propagation

In lower plants

(*i*) By budding

Experiment: Some yeasts are grown in sugar solution and observed under the microscope from time to time.

Observation: It is observed that one or more tiny outgrowths appear on one or more sides of the vegetative cells immersed in sugar solution. In some cases the outgrowths may be detached from the mother cell (which will grow into new individuals).

INFERENCE: This method of outgrowth formation is known as budding. Often budding continues one after the other so that finally a chain of cells is formed. All the individual cells of the chain separate from one another and form new yeast plants.

(*ii*) BY GEMMAE

EXPERIMENT: Some gemmae cups are collected from *Marchantia* and placed in its natural habitat.

OBSERVATION: A new thallus of *Marchantia* is grown from a gemma cup.

INFERENCE: This is a special method of vegetative reproduction. The gemmae develop in the gemmae cup and each gemma is a small, more or less circular flattened structure with a conspicuous depression on each side. The growing point lies in the depression.

(*iii*) BY LEAF TIP

EXPERIMENT: The leaf tip of *Adiantum* is made to touch the soil in which it grows and after a few days observation is made.

OBSERVATION and INFERENCE: When the leaf bends down and touches the ground the tip strikes root and forms a bud. The bud grows into a new independent fern plant.

IN HIGHER PLANT

(*iv*) BY UNDERGROUND STEMS

EXPERIMENT: Some tubers of potato or bulbs of onion or rhizomes of ginger are planted in pots. Observation is made after 10 to 15 days.

OBSERVATION: Many flowering plants reproduce themselves by means of rhizome, tuber, bulb or the corm. New buds are produced on these modified stems which gradually grow up into new plants.

(*v*) BY SUBAERIAL STEMS

EXPERIMENT: Some *Pistia*, *Hydrocotyle*, *Colocasia* or *Chrysanthemum* plants are observed.

OBSERVATION AND INFERENCE: Here vegetative propagation takes place by means of subaerial stems, e.g., by runner in *Hydrocotyle*, by offset in *Pistia*, y stolon in *Colocasia* and by sucker in *Chrysanthemum*.

(*vi*) BY ADVENTITIOUS BUDS

EXPERIMENT: Some *Bryophyllum* leaves are kept in moist soil.

OBSERVATION AND INFERENCE: A series of adventitious foliar buds are produced on the leaf margin (notch), each at the end of vein. These buds grow up into new plants.

(*vii*) BY BULBILS

EXPERIMENT: In *Globba bulbifera* or *Allium sativum* some of the lower flowers of the inflorescence become modified into small multicellular bodies, known as bulbils (Figure 44). Some bulbils are kept on moist ground and observed after a few days.

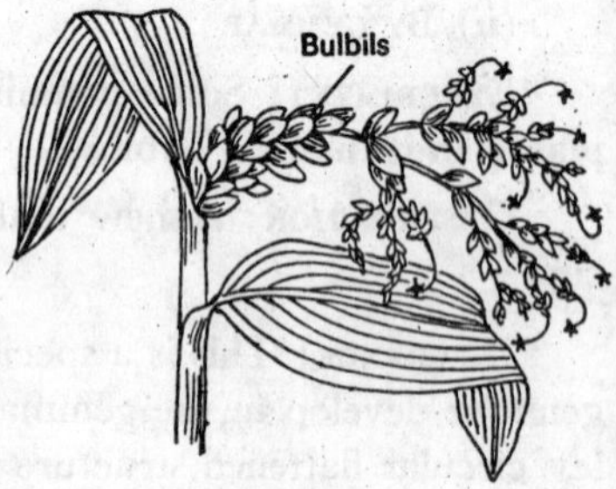

Fig. 44.

OBSERVATION AND INFERENCE: Bulbils grow up into new plants (sometimes they are seen to grow to some extent on the plant itself).

B. ARTIFICIAL METHODS OF PROPAGATION

(*i*) BY CUTTINGS

(*a*) BY STEM CUTTINGS

When some stem cuttings of rose, china-rose, *Moringa*, *Coleus*, etc., are put (the physiological polarity should be maintained) into moist soil they strike roots at the base and develop adventitious buds which grow up.

N.B. The rooting of the stems may be hastened by treatment of the cuttings with growth hormones like IBA, NAA, NOA, etc. in lanolin paste or in solution.

(*b*) BY ROOT CUTTINGS

When some root cuttings of lemon, *Citrus*, tamarind, etc. are put into moist soil they sprout forming roots and shoots.

(*ii*) BY LAYERING

In this case a lower branch of lemon, *Ixora*, rose, jesmine, etc., is bent down, a ring of bark to the length of 1 to 2 inches are removed and this portion is pushed into the soft ground keeping the upper part free. The bent portion is covered with soil and a stone or a brick is placed on it (Figure 45). When

Fig. 45.

roots have developed (usually within two to four months), the branch is cut out from the mother plant and grown separately.

(*iii*) BY GOOTEE

This method is usually employed for propagating lemon, orange, guava, litchi, *Magnolia*, etc. During early rains a healthy, somewhat woody branch is selected and a ring of bark (1 to 2 inches in length) is sliced off from it. A sufficiently thick plaster of grafting clay (clay two parts, cowdung one part and some finely-cut hay mixed with water) is applied all round the ringed portion which is then wrapped up with straw (or polythene bag) and tied in a secured manner (Figure 46). It should be wetted with water every morning and afternoon or may be kept moistened with the help of a wick from a water reservoir. Usually within one to three months the gootee is ready as is indicated by its striking roots. It is then cut out below the bandage and grown separately.

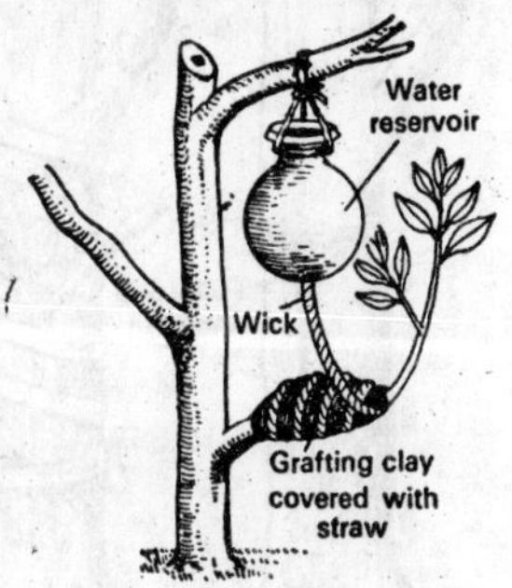

Fig. 46.

(*iv*) BY GRAFTING

(*a*) INARCHING OR APPROACH GRAFTING

By this method a branch (scion) of a plant is made to unite with a seedling (stock) by firmly tieing them together by means of a chord (Figure 47). Before doing this, a small portion of the bark is sliced off from each to ensure a closed contact and quicker union between the two. When proper fusion has taken place (usually within two to three months), the stock is cut out above the joining and the scion below, thus leaving the scion standing on the stock. Some of the fruit trees like mango, litchi, guava, plum, etc., readily respond to this method.

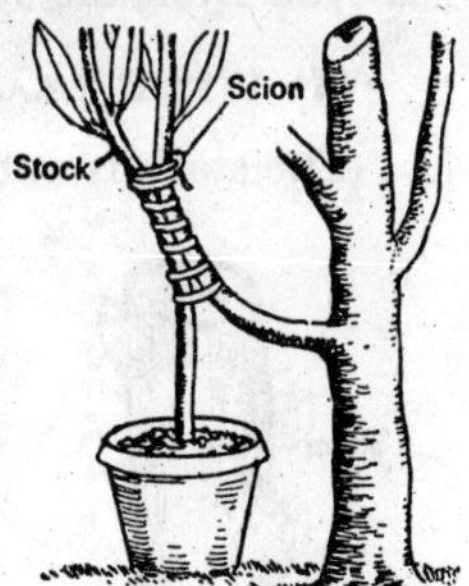

Fig. 47.

(*b*) BUD GRAFTING

For this method a T-shaped (Figure 48) incision is made in the bark of the stock, and a bud, cut out clean from a selected plant, is inserted into the T-shaped slit and properly bandaged. By this method it has been found possible to grow several varieties of roses on one rose stock, good varieties of orange, lemon, etc., on inferior stocks, several varieties of china-rose on one, cacti on one and so on.

(*c*) Whip or tongue grafting

The stock, usually half to three-fourth inch thick, is cut down a few inches above the ground, sloping cuts are then made in it 2 or 3 inches long, as shown in Figure 49. Scion of the same thickness is also cut in such a way

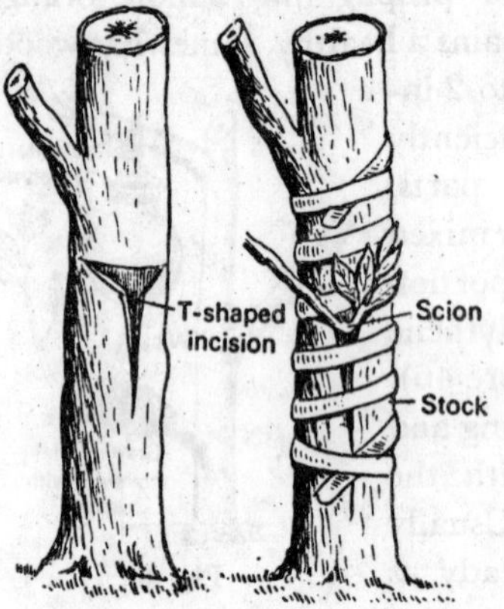

Fig. 48.

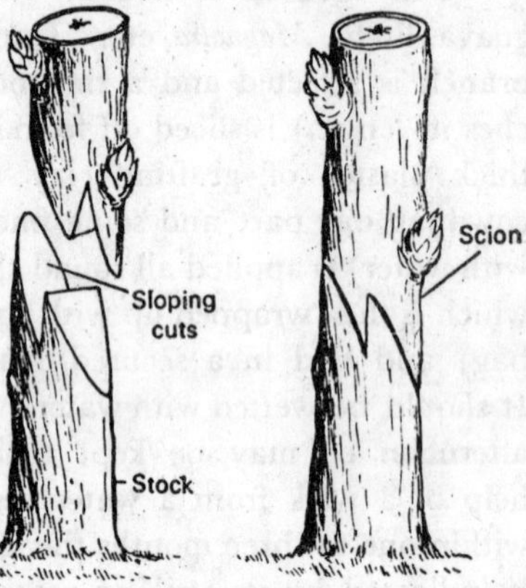

Fig. 49.

as to fit exactly into the stock. It is then inserted into the stock and tied firmly. The wound is of course covered with grafting wax (a mixture of tallow (animal fat) one part plus bees wax one part plus resin four parts, melted together and made into a soft dough under water). All buds are removed from the stock but not from the scion.

(*d*) Wedge grafting

The stock is cut 8 to 10 inches above the ground and the wood of the

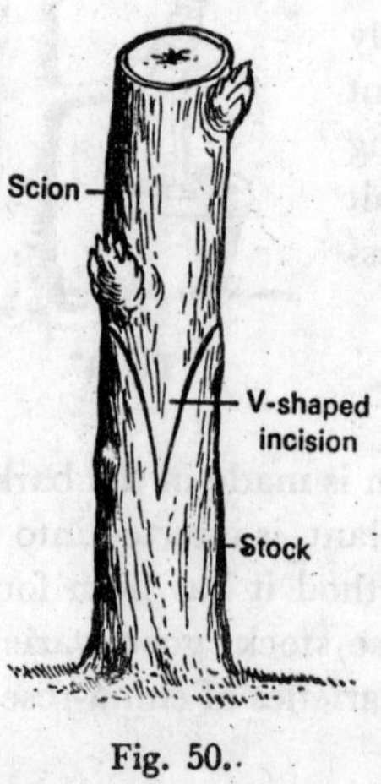

Fig. 50.

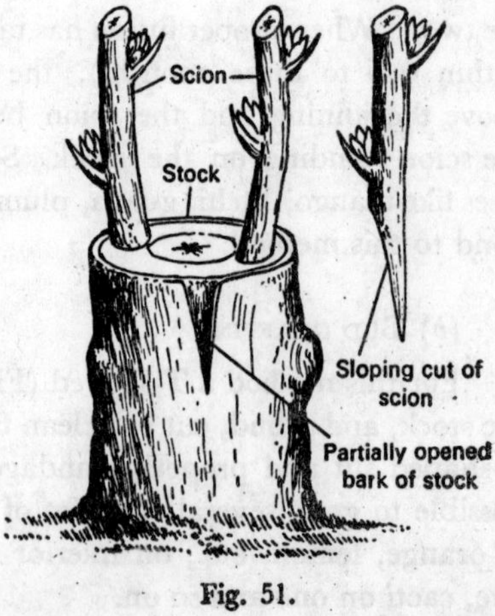

Fig. 51.

stem incised with clean cut in the form of a 'V'. The scion, cut obliquely downward so as to closely fit into the stock, is inserted into the stock and tied firmly (Figure 50). Grafting wax is used for covering the wound. All buds are removed from the stock but not from the scion.

(*e*) Crown grafting

An old tree may be rejuvenated by this method. The stem is cut across 8 to 10 inches above the ground. The bark of the stock is cut through from the surface downward to a length of 5 to 6 inches.. The bark is partially opened on either side. Prior to this a small branch cut out from a tree is incised at the base with a sloping cut and this is now inserted into the slit in the bark and tied firmly. The wound is covered with grafting wax (Figure 51).

2. Asexual reproduction

A. By fission

In many unicellular algae, fungi and bacteria, the mother cell splits into two new cells. The new cells, thus formed, contain all materials of the mother cell and soon they grow to the size of the latter, becoming a new independent plant. This process of fission may be observed in case of yeast cells under the microscope.

B. By spore formation

(*i*) Motile spores

Zoospores of algae like *Ulothrix*, *Vaucheria*, etc., may be observed under the microscope. The zoospores after escape from the mother cell germinate giving rise to new filaments.

(*ii*) Non-motile spores

Non-motile spores of some algae, e.g., *Nostoc*, *Penicillium*, *Phytophthora*, etc., are well adapted for dispersal by wind and at the same time to meet the ever-changing conditions of the environment. In favourable conditions they germinate and give rise to new plants.

(*iii*) True spores

True spores are always borne by sporophytes of bryophyta and pteridophyta. The germinatian of these spores may be observed in the laboratory under suitable culture conditions.

3. Sexual reproduction

A. By conjugation

In lower algae and fungi the pairing gametes are not differentiated into male and female (*isogamous*). The union of such similar gametes (*conjugation*) forms the zygote called *zygospore*. Conjugation of *Spirogyra* or *Mucor* may be observed under the microscope. Both scalariform and lateral conjugation may be observed in *Spirogyra* and only scalariform in case of *Mucor*.

B. By fertilisation

In higher plants the uniting gametes are differentiated into male and female and the union of the two (*fertilisation*) forms the zygote (*oospore*). When some pollen grains are dusted on the stigma of the flower, the pollen grains germinate forming a tube like outgrowth called pollen tube. The pollen tube contains sperms which unite with the egg of the nucellus. The pollen tube grows through the style and ultimately reaches the ovule. The growth of the pollen tube may be studied by cutting and staining the serial longitudinal sections of the style at different time intervals.

4. Induced sex modification

A. By growth hormones

IAA and other synthetic growth substances like 2, 4–D or NAA seem to be effective in the induction and modification of sex organs.

(*i*) Conversion of male flowers into female flowers by NAA

In dioecious species of *Cannabis sativa* genetically male plant could be induced to produce female flowers if during the period of differentiation of flower buds the third and fourth leaves of male plant are treated with 50 ppm NAA and also by 900 ppm ethrel. This indicates that the sexuality in this plant is determined by the concentration of native auxin and ethylene during the period of primordium differentiation and that femaleness is associated with a relatively high auxin level.

(*ii*) Increased production of female flowers by NAA

In some monoecious cucumber plant painting of the lower surface of the leaf with 50 ppm NAA causes an increase in the proportion of female flowers, some times altogether suppressing the production of male flowers. This also indicates that the female flowers tend to differentiate under higher concentration of auxins than do male flowers.

(*iii*) Promotion of maleness in sex expression by GA_3

Gibberellin increases the staminate flowers in some cases, e.g., cucumber in contrast to promotion of femaleness by auxin.

B. By growth retardants

Growth retardants also modify sex expression. The growth retardants are not the causative agents of sex expression in flowers. But the influence they exert may be due to differently regulated vegetative growth, and antagonistic reactions with growth promoters.

Inhibition of staminate flowers

The synthetic growth retardants such as CCC (2-Chloroethyl trimethyl ammonium chloride) and its derivatives MAB (Methyl ammonium bromide) and BCB (2-Bromoethyl trimethyl ammonium bromide) and Ethrel (2-Chloroethyl phosphonic acid) all these inhibit staminate flower production and greatly increase pistillate flower formation in cucumber and spinach. The greatest increase in pistillate flower formation is caused by MAB under high light intensities.

C. By other methods

(i) By carbon monoxide (CO) gas

In some Euphorbiaceae CO gas in low concentration some times reduces the number of male flowers in genetically monoecious types. The effect is presumably due to the effect of CO on auxin.

(ii) By salinity

Increase in salinity of the medium, particularly NaCl, may change the expression of sex in plants. In dioecious species, salinity may some times significantly cause an increase in the production of staminate flowers.

(iii) By photoperiod treatment

The length of photoperiod also affects the differentiation of sexes. If seeds of some Euphorbiaceae are grown in short days instead of long days (in long days the plants are predominantly staminate) the percentage of pistillate plants significantly increases. Similar results are also obtained with unisexual plants of *Ambrosia* of compositae. *Cannabis sativa* growing under 16 hours photoperiod produces about half staminate and half pistillate plants. But if the day length is shortened to 8 hours the plants are about half bisexual, half females and none males. Long days seem to favour production of male sex organs.

APPENDIX I

SIMPLE STATISTICAL CALCULATIONS

Statistics is that branch of applied mathematics which deals with collection of numerical observations and their interpretations in logical terms. When statistical observations and interpretations deal with living organisms or any biological problems, this science is called *biometry*. The main object of this science is to find out a reasonable conclusion in the presence of wide range of variations in both organisms and their environments. This also provides a measure for the probable truth of statements which are applicable to the whole group of individuals or any individual of a group of species.

Before going to interpret some of the more important statistical formulae which are commonly used in plant physiological problems, it is desirable to get oneself acquainted with some terms of statistics which one often comes across in such problems.

STATISTICAL METHODS : That deal with interpretation of observations in mathematical terms.

DESIGN OF EXPERIMENTS : The part dealing with the collection of observations.

QUALITATIVE AND QUANTITATIVE CHARACTERS : The former deals with the quality or attribute of individuals such as colour of flowers, surface nature of seeds, etc. and the latter deals with quantitative characters distinguished by measurements or counts such as height of a plant, yield of a crop, etc. In quantitative characters the variation may be *continuous* or *discrete*. A quantity that varies from individual to individual is called *variate* and the aggregate of individual values or the individuals themselves from which these are obtained is called a *population*.

FREQUENCY : The number of qualitative or quantitative characters occurring in each class.

FREQUENCY DISTRIBUTION : The manner in which the frequencies are distributed over different classes.

CLASS VALUE : The *mid-point* of each class (i.e., mean value of observations within a class) is termed as *class value*. As for example, if the shortest earhead of a wheat variety is 5·3 cm and the longest one is 13·7 cm, the range is 5·3 to 13·7 cm and this can be conveniently divided into 17 classes by taking class intervals of 0·5 cm. The frequency of the class consists of the number of observations falling within the limits of that class.

Some commonly used statistical calculations

Mean : It is the arithmetic average and is the result obtained when the sum of the values of individuals in the data is divided by the number of individuals in the data. The mean is usually denoted by the symbol M and is given by the formula

$M = \frac{\Sigma x}{N}$, where Σ represents the summation which is taken over all the N observations in the population.

When the data are expressed in a frequency distribution, the mean is calculated from the formula :

$$M = \frac{\Sigma f \cdot x}{N},$$

where f = frequency in the class,
x = class value,
N = total number of observations,
Σ = summation (of the products f with x over all classes).

Example 1. Mean

(*a*) When the data are directly taken into consideration:

Determination of the mean height of a plant population

Number of observations	*Height* (cm)
1	24·7
2	22·9
3	28·4
4	27·5
5	23·4
6	22·9
7	19·8
8	26·3
9	25·6
10	22·5
N=10	Σx=244·0

Hence, $M = \frac{\Sigma x}{N} = \frac{244}{10} = 24{\cdot}4$ cm.

(*b*) When the data are distributed into classes in case of large populations :

Determination of the mean height of 200 plants

Class value (x) (cm)	*Frequency* (f)	*Class value* × *Frequency* ($f.x$)
40	28	1120
42	30	1260
44	8	352
46	12	552
48	50	2400
50	5	250
52	67	3484
	N=200	$\Sigma f.x$=9418

Hence, $M = \frac{\Sigma f \cdot x}{N} = \frac{9418}{200} = 47{\cdot}09$ cm.

Mean deviation : This is one of the measures of *dispersion* or *spread*. It is calculated by adding the deviations of all individual observations from their arithmetic mean without regard to the sign and dividing the sum by the total number of observations. The following formula is applied for its determination.

$$M.\ D.\ (\text{mean deviation}) = \frac{\Sigma f.d}{N},$$

where f = class frequency,

d = deviation of the class value from the mean, always taken positive $(x-M)$,

N = total number of observations and

Σ = summation taken over all the classes of the distribution.

Standard deviation : The measure of dispersion or spread or variability of a sample around the mean is known as the *standard deviation.* It is calculated by squaring the deviation of each observation from the mean, adding the squares, dividing by the number of observations and extracting the square root according to the formula :

$$S.\ D. = \sqrt{\left(\frac{\Sigma d^2}{N-1}\right)}$$

where $S.D.$ stands for standard deviation and where summation is taken over all N observations. When the data are grouped in the form of a frequency distribution, the formula takes the form :

$$S.\ D. = \sqrt{\left(\frac{\Sigma f.d^2}{N-1}\right)}$$

Since the deviations, both positive and negative, are squared, the products $f.d^2$ are always positive.

N.B. Here instead of N, $(N-1)$ has been used and this means the *number of degrees of freedom* and it denotes the number of comparisons that can be made between any observation and the rest of the observations taking them in pairs, since only a single observation can be compared with each of the remaining $(N-1)$ taking one at a time.

Though the standard deviation may be calculated from the above formula, but a correction factor has to be applied to compensate for the error introduced by grouping the observations into classes and basing the calculations on the values of class centres. This correction is called *Sheppard's correction.* The Sheppard's correction is (1/12) of the square of the class interval and is to be deducted from $\sqrt{\left(\frac{\Sigma f.d^2}{N-1}\right)}$.

This will be clear from the following example :

EXAMPLE 2. Calculation of standard deviation of number of grains per earhead in 12 wheat plants.

Class	*Class value* (x)	*Frequency* (f)	*Mean* (M)	*Deviation from the mean* (d)	*Deviation squared* (d^2)	*Frequency × Deviation squared* ($f.d^2$)
8–12	10	1		−20·55	422·3	422
13–17	15	17		−15·55	241·8	4111
18–22	20	25		−10·55	111·3	2782
23–27	25	86		−5·55	30·8	2649
28–32	30	125	30·55	−0·55	0·3	·38
33–37	35	77		4·45	19·8	1525
38–42	40	55		9·45	89·3	4912
43–47	45	9		14·45	208·8	1879
48–52	50	4		19·45	378·3	1513
53–57	55	1		24·45	597·8	598
Total	..	N=400				$\Sigma f.d^2$=20429

$$S.\,D. = \sqrt{\frac{\Sigma f\,.\,d^2}{N} - \tfrac{1}{12}(I)^2},$$ where I is the class interval

$$= \sqrt{\frac{20429}{400} - \tfrac{1}{12}(5)^2}$$

$$= \sqrt{(51{\cdot}0725 - 2{\cdot}0833)}$$

$$= 6{\cdot}999.$$

Without Sheppard's correction, the value of *S. D.* would have been 7·146.

Variance : It is a measure of variation and is used to denote the square of the standard deviation, i.e., $(S.\,D.)^2$. Thus

$$Variance = (S.\,D.)^2 = \frac{\Sigma f\,.\,d^2}{N-1}.$$

EXAMPLE 3. Standard deviation

DETERMINATION OF STANDARD DEVIATION IN HEIGHT OF A SAMPLE

(*a*) When the data are directly taken into consideration:

From the data of the *Example* 1(*a*) the measurements are tabulated and calculated as follows.

Number of observations	*Height* (cm) ($\bar{M}$)	*Mean* (M)	*Deviation from mean* ($\bar{M}-M$) (d)	*Deviation squared* (d^2)
1	24·7		0·3	0·09
2	22·9		−1·5	2·25
3	23·9		−1·0	1·00
4	27·5		3·1	9·61
5	23·4		−1·0	1·00
6	22·9	24·4	−1·5	2·25
7	19·8		−4·6	21·16
8	26·3		1·9	3·61
9	25·6		1·2	1·44
10	22·5		−1·9	3·61
N=10				Σd^2=61·02

Hence, $S.\,D. = \sqrt{\left(\frac{\Sigma d^2}{N-1}\right)} = \sqrt{\left(\frac{61{\cdot}02}{9}\right)} = 2{\cdot}60.$

The mean is thus expressed as (24·4±2·60) cm.

(*b*) When the data are distributed into classes :

From the data of *Example* 1(*b*) the measurements are tabulated and calculated as follows:

Class value (x)	*Frequency* (f)	*Mean* (M)	*Deviation from mean* $\{d=(x-M)\}$	*Deviation squared* (d^2)	*Frequency* × *Deviation squared* ($f \times d^2$)
40	28		7·08	50·12	1403·36
42	30		5·08	25·80	774·00
44	8		3·08	9·48	75·84
46	12	47·08	1·08	1·16	13·92
48	50		−0·92	0·846	42·30
50	5		−2·92	8·52	42·60
52	67		−4·92	24·20	1621·40
	N=200				$\Sigma f.d^2$=3973·42

Hence, $S.\,D. = \sqrt{\left(\frac{\Sigma f\,.\,d^2}{N-1}\right)} = \sqrt{\left(\frac{3973{\cdot}42}{199}\right)} = \sqrt{19{\cdot}96} = 4{\cdot}46.$

The mean therefore can be expressed as (47·08 ± 4·46) cm.

Variance = $(S.\,D.)^2$ = 19·96.

Coefficient of variation : This is a measure of dispersion or spread or variation which is independent of the unit of measurement and is therefore useful for comparison between different populations. This measure is known as the *coefficient of variation* which is the standard deviation expressed as percentage of the mean. This is calculated as follows:

$$C.V. = \frac{S.D.}{M} \times 100$$

EXAMPLE 4. From the *Examples* 1(*b*) and 3(*b*) we can calculate the coefficient of variation in that sample.

$M = 47{\cdot}09$ cm; $S.D. = 4{\cdot}46$

Hence, $C.V. = \dfrac{4{\cdot}46 \times 100}{47{\cdot}09} = 9{\cdot}47$ per cent.

Standard error : This is a measure of the variation of the means and hence the standard deviation of the mean is commonly termed the *standard error*. If *S.E.* represents the standard deviation of the mean (i.e., standard error) of a sample of *N* individuals and *S.D.* stands for the deviation of the parent population then,

$$S.E. = \frac{S.D.}{\sqrt{N}}$$

The square of the standard error is called the *sampling variance.*

EXAMPLE 5. Standard error determination.

(*a*) From the Example 3(*a*) we get $S.D. = 2{\cdot}60$ and $N = 10$

Hence, $S.E. = \dfrac{S.D.}{\sqrt{N}} = \dfrac{2{\cdot}60}{\sqrt{10}} = \dfrac{2{\cdot}60}{3{\cdot}16} = 0{\cdot}82.$

(*b*) From the Example 3(*b*) we get $S.D. = 4{\cdot}46$ and $N = 200$

Hence, $S.E. = \dfrac{4{\cdot}46}{\sqrt{200}} = \dfrac{4{\cdot}46}{14{\cdot}14} = 0{\cdot}315.$

Thus the mean and the standard error can be expressed as $(47{\cdot}08 \pm 0{\cdot}315)$ cm.

Significance test : It is the calculation of probability of obtaining an observed result from some hypothesis and whether the hypothesis is to be rejected (if insignificant) or accepted (if significant) can be known from the significance test.

'*t*' TEST : It is one of the simplest and commonly used test for significance. It is a quantity representing the difference between the sample mean and the true mean or population mean expressed in terms of standard error.

$$t = \frac{\text{Mean}}{\text{Standard error}} = \frac{M - \mu}{S.E.}$$

where *M* is the sample mean,

μ is the population mean (a hypothetical value usually regarded as 0) and

S.E. is the standard error of the sample mean.

When two samples of small size are compared, the formula for t becomes

$$t = \frac{M_A - M_B}{S.E._d}$$

where M_A and M_B represent the means of the samples A and B,

$S.E._d$ is the standard error of difference in means and can be calculated as follows :

$$S.E._d = \sqrt{\{(S.E._A)^2 + (S.E._B)^2\}}$$

where $S.E._A$ and $S.E._B$ are standard errors of samples A and B respectively.

Now the calculated value of t is compared with that of t in the table for distribution of t (Table 1) against the degrees of freedom $(N-1)$, at $P = 0{\cdot}05$, $0{\cdot}02$ and $0{\cdot}01$. If the calculated value of t is found to be greater than the value from the table, then the observed difference is considered to be statistically significant and if, on the other hand, the calculated value of t is less than the table value of t then the difference is not significant. Usually the 5% level is considered for significance test (i.e., at $P=0{\cdot}05$) for ordinary purposes.

N.B. If the sample number is small (less than 30) the mean and the standard deviation can be directly calculated from the population, instead of putting them into classes. Here the value of N is always equal to the total number of population.

Let us compare the means of the results of A and B series. The same will be applicable for B and C or A and C. This means that we would like to know whether A and B values significantly vary or not. In order to do that the popular 'Student's t-test' can be applied.

Step 1. Calculate the standard error of difference between the two means A and B, i.e., $S.E._A - S.E._B$.

Step 2. Calculate the value of t

$$t = \frac{\text{Difference between the arithmetic means}}{\text{Standard error of difference}}$$

$$= \frac{A-B}{S.E._A - S.E._B}$$

Step 3. Now consult the t-table from the standard table against the sum of the degrees of freedom for each series A and B and at 5% level of significance. If the calculated value is higher than that given in the standard table the difference is significant. A short-cut method is that if the value is greater than 2 the difference is significant.

EXAMPLE 6. Significance test (t-test)

Significance test of an observed value from hypothetical value.

(*a*) WHEN THE DATA ARE DIRECTLY TAKEN INTO CONSIDERATION

From the data of the Example 5(a) we get $S.E. = 0{\cdot}82$.

$M = 47{\cdot}08$ and μ = a hypothetical value say 45.

$$\text{Hence, } t = \frac{M-\mu}{S.E.} = \frac{47{\cdot}08-45}{0{\cdot}82} = \frac{2{\cdot}08}{0{\cdot}82} = 2{\cdot}53.$$

Here t at 5% level has been found to be significant against 9 degrees of freedom.

(*b*) COMPARISON OF DIFFERENCE BETWEEN TWO SAMPLES BY *t*-TEST

Effect of NAA and IAA on height of a plant population :

Treatment with NAA

No. of Obs.	*Height* (cm)	*Mean* (cm)	*d*	d^2
1	23·4		2·1	4·41
2	20·2		—1·1	1·21
3	23·3		2·0	4·00
4	24·8		3·5	12·25
5	21·8	21·3	0·5	0·25
6	22·8		1·5	2·25
7	17·5		—3·8	14·44
8	15·6		—5·7	32·49
9	24·4		3·1	9·61
10	19·2		—2·1	4·41
N=10	Σx =213			Σd^2=85·32

$$S.E._{NAA} = \frac{\sqrt{\left(\frac{\Sigma d^2}{N-1}\right)}}{\sqrt{N}} = \frac{\sqrt{\left(\frac{85{\cdot}32}{9}\right)}}{\sqrt{10}} = \frac{3{\cdot}08}{3{\cdot}16} = 0{\cdot}974.$$

Treatment with IAA

No. of Obs.	*Height* (cm)	*Mean* (cm)	*d*	d^2
1	24·7		0·3	0·09
2	28·4		4·0	16·00
3	22·9		—1·5	2·25
4	27·5		3·1	9·61
5	23·4	24·4	—1·0	1·00
6	26·3		1·9	3·61
7	22·9		—1·5	2·25
8	19·8		—4·6	21·16
9	25·6		1·2	1·44
10	22·5		—1·9	3·61
N=10	Σx =244			Σd^2=61·02

$$S.E._{IAA} = \frac{\sqrt{\left(\frac{\Sigma d^2}{N-1}\right)}}{\sqrt{N}} = \frac{\sqrt{\left(\frac{61{\cdot}02}{9}\right)}}{\sqrt{10}} = \frac{2{\cdot}60}{3{\cdot}16} = 0{\cdot}823.$$

$$\text{Therefore, } t = \frac{M_{NAA} - M_{IAA}}{\sqrt{(S.E._{NAA})^2 + (S.E._{IAA})^2}}$$

$$= \frac{21{\cdot}3 - 24{\cdot}4}{\sqrt{\{(0{\cdot}974)^2 + (0{\cdot}823)^2\}}} = \frac{3{\cdot}1}{1{\cdot}275} = 2{\cdot}431 \text{ (negative sign may be ignored).}$$

Now the value of t at 5% level is found to be significant from the standard table.

Correlation and correlation coefficient : It often happens that changes in one variable are accompanied by changes in another and that a definite relation exists between the two. In other words, there is a *correlation* between the variables.

When two variables change together in such a way that an increase in one variable is accompanied by an increase in the other, the variables are said to be *positively correlated.* On the other hand, if an increase in one variable is accompanied by a decrease in the other, these two variables are said to be *negatively correlated.* If there is no relationship between the two variables they are said to be *independent* or *uncorrelated.*

The correlation coefficient (r) may be defined as a statistical measure of relationship between two or more variables.

$$\text{Correlation coefficient } (r) = \frac{\Sigma\{(x-\bar{x}) \times (y-\bar{y})\}}{\sqrt{\{\Sigma(x-\bar{x})^2 \times \Sigma(y-\bar{y})^2\}}}$$

where x and y represent the two variables; $(x-\bar{x})$ and $(y-\bar{y})$ are deviations from the means (ref. variance and $S.D.$).

This equation can be shortened as follows :

$$r = \frac{\Sigma dx \times dy}{\sqrt{(\Sigma d^2x \times \Sigma d^2y)}}$$

where, dx and dy denote deviations in the variables x and y respectively.

EXAMPLE 7. Correlation coefficient of length and weight of seeds of a species.

No. of Obs.	*Length* (cm) x	*Weight* (gm) y	x^2	y^2	$x \times y$
1	2·1	1·0	4·41	1·00	2·1
2	2·0	1·0	4·00	1·00	2·0
3	1·5	0·9	2·25	0·81	1·3
4	1·9	0·9	3·61	0·81	1·7
5	1·7	0·6	2·89	0·36	1·0
6	2·3	1·4	5·29	1·96	3·2
7	2·1	1·0	4·41	1·00	2·1
8	2·2	1·1	4·84	1·21	2·4
9	2·4	1·2	5·76	1·44	2·8
10	2·2	1·0	4·84	1·00	2·2
N=10	Σx=20·40	Σy=10·10	Σx^2=42·30	Σy^2=10·59	Σxy=21·0

Now the values of $\Sigma dx.dy$, $\Sigma d^2 x$ and $\Sigma d^2 y$ are calculated as follows :

$$\Sigma dx.dy = \Sigma x.y - \frac{(\Sigma x \times \Sigma y)}{N}$$

$$= 21{\cdot}00 - \frac{20{\cdot}4 \times 10{\cdot}1}{10} = 21{\cdot}00 - 20{\cdot}60 = 0{\cdot}40$$

$$\Sigma d^2 x = \Sigma x^2 - \frac{(\Sigma x)^2}{N} = 42{\cdot}30 - \frac{(20{\cdot}4)^2}{10} = 42{\cdot}30 - 41{\cdot}61 = 0{\cdot}69$$

$$\Sigma d^2 y = \Sigma y^2 - \frac{(\Sigma y)^2}{N} = 10{\cdot}59 - \frac{(10{\cdot}1)^2}{10} = 10{\cdot}59 - 10{\cdot}20 = 0{\cdot}39$$

Therefore, $r = \frac{0{\cdot}40}{\sqrt{(0{\cdot}69 \times 0{\cdot}39)}} = \frac{0{\cdot}40}{\sqrt{0{\cdot}2691}} = \frac{0{\cdot}40}{0{\cdot}518} = 0{\cdot}78$

Now from the Table of r check up the significance of the calculated value against 9 degrees of freedom at 5% level ($P=0{\cdot}05$). If the calculated value is higher than the table value, the result is significant. In the above example the value is found to be significant. Here the correlation is also positive since one variable increases with the other. When r is positive (say, plus 1), the correlation is positive and perfect and when r is negative (say, minus 1), the correlation is negative and perfect.

Least significant difference (*L.S.D.*): The *L.S.D.* provides an easy way to distinguish between differences which are not due to chance and those that are. The test is made by computing a difference which is significant at the 5% level and then comparing it with the sample differences in the experiment. *L.S.D.* is calculated as follows :

1. Design of the experiment : Random block design:

Example :

Treatments	*Replications*				
	I	II	III	*Total*	*Mean*
A	1	2	3	6	2
B	4	5	6	15	5
C	7	8	9	24	8
Total	12	15	18	45 (Trt. totals)	
		(Repl. totals)			

1. Find out the Correction Factor (*C. F.*) : It is calculated from the above table.

$$C.\,F. = (Sx)^2/N$$

ere Sx is the total of data and N is the number of observations. Thus from the above table,

$$C.\,F. = \frac{(45)^2}{9} = 2025/9 = 225.$$

2. Calculate Sums of Squares (*S.S.*) for each source of variation :

Total $S.S. = (1^2+2^2+3^2+ \dots +8^2+9^2) - C.F.$

$= (1+4+9+16+25+36+49+64+81) - 225$

$= 285 - 225 = 60.$

3. Calculate Sums of Squares for Replication (*R.S.S.*) :

$$R.S.S. = \frac{(12^2+15^2+18^2)}{\text{No. of treatments}} - C.F. = \frac{(144+225+324)}{3} - C.F.$$

$$= \frac{688}{3} - C.F. = 229{\cdot}3 - 225 = 4{\cdot}3.$$

4. Calculate Sums of Squares for Treatment (*T.S.S.*) :

$$T.S.S. = \frac{(6^2+15^2+24^2)}{\text{No. of replications}} - C.F. = \frac{36+225+576}{3} - C.F.$$

$$= \frac{837}{3} - C.F. = 279 - 225 = 54.$$

5. Calculate Sums of Squares for Error (*E.S.S.*) :

$E.S.S. =$ Total $S.S. - (R.S.S. + T.S.S.)$

$= 60 - (4{\cdot}3+54) = 60 - 58{\cdot}3 = 1{\cdot}7.$

6. The four sums of squares calculated above are entered in the table of 'Analysis of Variance' :

Source of variation	*Degrees of Freedom*		*Sums of squares*	*Mean squares*	*F-value*
	Symbolic	*Numerical*			
Replication (r)	$r-1$	2	4·3	$\frac{4{\cdot}3}{t-1}$ $=4{\cdot}3/2$ $=2{\cdot}15$	$\frac{M.S.\ \text{Repl.}}{M.S.\ \text{Error}}$ $=2{\cdot}15/0{\cdot}425$ $=5{\cdot}3$ (Not significant)
Treatment (t)	$t-1$	2	54	$\frac{54}{r-1}$ $=54/2$ $=27$	$\frac{M.S.\ \text{Trt.}}{M.S.\ \text{Error}}$ $=27/0{\cdot}425$ $=63{\cdot}5^*$ (significant)
Error	$(r-1)(t-1)$	4	1·7	$\frac{1{\cdot}7}{(r-1)(t-1)}$ $=1{\cdot}7/4$ $=0{\cdot}425$	
Total	$r.t-1$	8	60		

N.B. If the *F*-value at replication level is ***significant*** then the experiment may be rejected for error due to small number of replications. When the

F-value at treatment level is significant, *L.S.D.* at 5% level may be calculated. But if this value is *not significant*, *L.S.D.* need not be calculated.

Find out the *F*-value for treatment level at degrees of freedom of error level (n_2) and treatment level (n_1); that for replication level at degrees of freedom of error level (n_2) and replication level (n_1) from Fisher's table at 5%.

7. Calculate *L.S.D.* at 5% level from the formula :

L.S.D. 5% $= t \times 1{\cdot}414 \times S\bar{x}$

where the value of *t* may be obtained from the *t*-Table at degrees of freedom for error term in the 'Analysis of Variance Table'.

$$S\bar{x} = \sqrt{\left(\frac{M.\ S.\ \text{for Error}}{\text{No. of Replications}}\right)}$$

$$= \sqrt{\left(\frac{0{\cdot}425}{3}\right)} = \sqrt{0{\cdot}1416} = 0{\cdot}376$$

Now, *L.S.D.* 5% $= t \times 1{\cdot}414 \times S\bar{x}$

$= 2{\cdot}78 \times 1{\cdot}414 \times 0{\cdot}376$

$= 1{\cdot}5$

Thus a difference of 1·5 and above between treatments will always be significant statistically.

Table 1. VALUES OF *t* FOR $P = 0{\cdot}05$, $0{\cdot}02$ AND $0{\cdot}01$.

Degrees of freedom (*n*)	*P*=0·05 (5%)	*P*=0·02 (2%)	*P*=0·01 (1%)
1	12·706	31·821	63·657
2	4·303	6·965	9·925
3	3·182	4·541	5·841
4	2·776	3·747	4·604
5	2·571	3·365	4·032
6	2·447	3·143	3·707
7	2·365	2·998	3·499
8	2·306	2·896	3·355
9	2·262	2·821	3·250
10	2·228	2·764	3·169
11	2·201	2·718	3·106
12	2·179	2·681	3·055
13	2·160	2·650	3·012
14	2·145	2·624	2·977
15	2·131	2·602	2·947
16	2·120	2·583	2·921
17	2·110	2·567	2·898
18	2·101	2·552	2·878
19	2·093	2·539	2·861
20	2·086	2·528	2·845
21	2·080	2·518	2·831
22	2·074	2·508	2·819
23	2·069	2·500	2·807

Degrees of freedom (*n*)	*P*=0·05 (5%)	*P*=0·02 (2%)	*P*=0·01 (1%)
24	2·064	2·492	2·797
25	2·060	2·485	2·787
26	2·056	2·479	2·779
27	2·052	2·473	2·771
28	2·048	2·467	2·763
29	2·045	2·462	2·756
30	2·042	2·457	2·750
40	2·021	2·423	2·704
60	2·000	2·390	2·660
120	1·980	2·358	2·617
∞	1·960	2·326	2·576

Table 2. VALUES OF *r* FOR DIFFERENT LEVELS OF SIGNIFICANCE.

Degrees of freedom (*n*)	*P*=0·05 (5%)	*P*=0·02 (2%)	*P*=0·01 (1%)
1	0·99692	0·999507	0·999877
2	0·95000	0·98000	0·99000
3	0·8783	0·93433	0·95873
4	0·8114	0·8822	0·91720
5	0·7545	0·8329	0·8745
6	0·7067	0·7887	0·8343
7	0·6664	0·7498	0·7977
8	0·6319	0·7155	0·7646
9	0·6021	0·6851	0·7348
10	0·5760	0·6581	0·7079
11	0·5529	0·6339	0·6835
12	0·5324	0·6120	0·6614
13	0·5139	0·5923	0·6411
14	0·4973	0·5742	0·6226
15	0·4821	0·5577	0·6055
16	0·4683	0·5425	0·5897
17	0.4555	0·5285	0·5751
18	0·4438	0·5155	0·5614
19	0·4329	0·5034	0·5487
20	0·4227	0·4921	0·5368
25	0·3809	0·4451	0·4869
30	0·3494	0·4093	0·4487
35	0·3246	0·3810	0·4182
40	0·3044	0·3578	0·3932
45	0·2875	0·3384	0·3721
50	0·2732	0·3218	0·3541
60	0·2500	0·2948	0·3248
70	0·2319	0·2737	0·3017
80	0·2172	0·2565	0·2830
90	0·2050	0·2422	0·2673
100	0·1946	0·2301	0·2540

Table 3. VALUES OF *F* (VARIANCE RATIO) AT 5% LEVEL

n_2 \ n_1	1	2	3	4	5	6	8	12	24	∞
1	161·40	199·50	215·70	224·60	230·20	234·00	238·90	243·90	249·00	254·30
2	18·51	19·00	19·16	19·25	19·30	19·33	19·37	19·41	19·45	19·50
3	10·13	9·55	9·28	9·12	9·01	8 94	8·84	8·74	8·64	8·53
4	7·71	6·94	6·59	6·39	6·26	6·16	6·04	5·91	5·77	5·63
5	6·61	5·79	5·41	5·19	5·05	4·95	4·82	4·68	4·53	4·36
6	5·99	5·14	4·76	4·53	4·39	4·28	4·15	4·00	3·84	3·67
7	5·59	4·74	4·35	4·12	3·97	3·87	3·73	3·57	3·41	3·23
8	5·32	4·46	4·07	3·84	3·69	3·58	3·44	3·28	3·12	2·93
9	5·12	4·26	3·86	3·63	3·48	3·37	3·23	3·07	2·90	2·71
10	4·96	4·10	3·71	3·48	3·33	3·22	3·07	2·91	2·74	2·54
11	4·84	3·98	3·59	3·36	3·20	3·09	2·95	2·79	2·61	2·40
12	4·75	3·88	3·49	3·26	3·11	3·00	2·85	2·69	2·50	2·30
13	4·67	3·80	3·41	3·18	3·02	2·92	2·77	2·60	2·42	2·21
14	4·60	3·74	3·34	3·11	2·96	2·85	2·70	2·53	2·35	2·13
15	4·54	3·68	3·29	3·06	2·90	2·79	2·64	2·48	2·29	2·07
16	4·49	3·63	3·24	3·01	2·85	2·74	2·59	2·42	2·24	2·01
17	4·45	3·59	3·20	2·96	2·81	2·70	2·55	2·38	2·19	1·96
18	4·41	3·55	3·16	2·93	2·77	2·66	2·51	2·34	2·15	1·92
19	4·38	3·52	3·13	2·90	2·74	2·63	2·48	2·31	2·11	1·88
20	4·35	3·49	3·10	2·87	2·71	2·60	2·45	2·28	2·08	1·84
21	4·32	3·47	3·07	2·84	2·68	2·57	2·42	2·25	2·05	1·81
22	4·30	3·44	3·05	2·82	2·66	2·55	2·40	2·23	2·03	1·78
23	4·28	3·42	3·03	2·80	2·64	2·53	2·38	2·20	2·00	1·76
24	4·26	3·40	3·01	2·78	2·62	2·51	2·36	2·18	1·98	1·73
25	4·24	3·38	2·99	2·76	2·60	2·49	2·34	2·16	1·96	1·71
26	4·22	3·37	2·98	2·74	2·59	2·47	2·32	2·15	1·95	1·69
27	4·21	3·35	2·96	2·73	2·57	2·46	2·30	2·13	1·93	1·67
28	4·20	3·34	2·95	2·71	2·56	2·44	2·29	2·12	1·91	1·65
29	4·18	3·33	2·93	2·70	2·54	2·43	2·28	2·10	1·90	1·64
30	4·17	3·32	2·92	2·69	2·53	2·42	2·27	2·09	1·89	1·62
40	4·03	3·23	2·84	2·61	2·45	2·34	2·18	2·00	1·79	1·51
60	4·00	3·16	2·76	2·52	2·37	2·25	2·10	1·92	1·70	1·39
120	3·92	3·07	2·68	2·45	2·29	2·17	2·02	1·83	1·61	1·25
∞	3·84	2·99	2·60	2·37	2·21	2·09	1·94	1·75	1·52	1·00

Lower 5% points are found by interchange of n_1 and n_2, i.e, n_1 must always correspond with the greater mean square.

Tables 1, 2 and 3 are taken from Tables III, V and VII of Fisher and Yates: *Statistical Tables for Biological, Agricultural and Medical Research*, published by Longman Group Ltd., London. (Previously published by Oliver & Boyd, Edinburgh), and by permission of the authors and publishers.

APPENDIX II

SOME USEFUL INFORMATIONS AND PREPARATION OF REAGENTS

Definitions of some physical and chemical terms

ATOMIC WEIGHT : Atomic weight of an element is the weight of an atom of the element relative to the weight of the oxygen atom taken as 16. On this basis, the atomic weight of hydrogen is 1·008. The *gram atomic weight* or *gram atom* of an element is its atomic weight expressed in grams : thus 16 grams of oxygen constitute 1 gm atom of oxygen. Atomic weight= Equivalent weight × Valency.

MOLECULAR WEIGHT : The molecular weight of a substance may be defined as the weight of a molecule of the substance relative to the weight of the oxygen atom as 16. Therefore the molecular weight of O_2 is 32. The molecular weight of a substance in grams is referred to as the *gram molecular weight* or the *gram molecule* or simply as a *mole.*

The gram molecular weight of any gas or vapour occupies 22·4 litres at N.T.P. (normal temperature and pressure). This is called *gram molecular volume* or simply *molar volume.*

AVOGADRO'S NUMBER : It is defined as the number of the individual molecules per gram molecule of a substance which is constant for all substances and its value is $6{\cdot}023 \times 10^{23}$.

EQUIVALENT WEIGHT : The equivalent weight of an element is defined as the number of parts by weight of that element which will combine chemically with or will replace from chemical combination 8 parts by weight of oxygen or its equivalent, i.e., 1·008 parts by weight of hydrogen or 35·5 parts by weight of chlorine. The equivalent weight is a number ; the *gram equivalent* is the equivalent weight expressed in gram. Equivalent weight=Atomic weight/Valency.

ATOMIC NUMBER : The electrons circling the nucleus are negatively charged and are exactly equal in number to the positively charged protons within the nucleus, thereby producing an electrically uncharged atom. All chemical and biological properties of the atom are due to the electrons, especially their number and the way their orbits around the nucleus are arranged. This number, which of course, is also the number of protons within the nucleus, is different for every element and determines its property. This is called the *atomic number* of that element. It ranges from 1 in hydrogen (one proton and one electron) to 100 and above for transuranium elements.

ISOTOPE : The atoms of an element having the same atomic number and similar chemical properties but having different atomic weights (atomic mass) are called isotopes of that element.

Isotopes are often designated by *mass number*, i.e., the total number of protons and neutrons in the nucleus of the atom. Thus carbon has three natural isotopes ^{12}C (6 protons plus 6 neutrons, mass number is 12), ^{13}C (6 protons plus 7 neutrons, mass number is 13) and ^{14}C (6 protons plus 8 neutrons, mass number is 14). There are two categories of isotopes—*stable* or *nonradioactive* and *unstable* or *radioactive* which always disintegrate by emitting radiations (alpha, beta or gamma rays). The *half-life* of a radioactive isotope is the time taken for the activity of a radioactive element to decay to half of its original value.

ABSOLUTE TEMPERATURE : According to Charle's law, at constant pressure the volume of a given mass of any gas expands by (1/273) of its volume at 0°C. for each 1°C. rise in temperature. If the gas were cooled to −273°C., it would have occupied no volume at that temperature, i.e., the volume becomes zero. This temperature is called the *absolute temperature* (T). The absolute temperature is obtained by taking −273°C. as zero degree absolute. The volume of a given mass of any gas at constant pressure is directly proportional to its absolute temperature.

DENSITY : Density is the mass per unit volume. $D=M/V$, where M is the mass, V the volume and D the density.

SPECIFIC GRAVITY : The relative density of a substance is the ratio of its density to the density of standard substance under identical condition of temperature and pressure. In the case of liquids and solids, the standard substance is water at 4°C.; the relative density is then called the *specific gravity*. In C.G.S. system, the absolute density of water at 4°C. is unity. Hence, the specific gravity of a substance is numerically equal to its density, for example, the absolute density of mercury is 13·6, the figure 13·6 also indicates its specific gravity.

ACIDS : Acids are substances which are capable of donating protons (hydrogen ions) and the extent of ionisation of an acid in a given solvent will depend on the capacity of the solvent for acceptance of the protons. Acids turn blue litmus red.

BASES : Bases are substances which are capable of combining with protons. Bases turn red litmus blue.

TITRATION : This is a process of volumetrically determining the unknown strength of a solution by reaction with a solution of known concentration (standard solution) in presence of an indicator.

NEUTRALISATION : In neutralisation reactions, acids and bases react with each other in equivalent amounts producing salt and water. When solutions of an acid and an alkali are mixed together in equivalent proportions, the resulting solution has neither acid nor alkaline character—this reaction is called *neutralisation*.

INDICATORS : Indicators are substances which are used to indicate, generally by their colour change, specific physico-chemical condition of chemical systems. There are three classes of indicators:

(1) *Acid-base indicators* like litmus, etc. used for acidimetric titrations.

(2) *Oxidation-reduction indicators* like diphenylamine used to indicate the end point of oxidation of ferrous ion by dichromate by a deep blue colour at the end point.

(3) *Absorption indicators* like eosin used in the titration of chloride with silver nitrate solution and its end point is indicated by red colour of the white silver chloride precipitate.

CHOICE OF INDICATORS : Choice of indicators depends upon their pH range. Different indicators indicate neutrality at different pH. In fact the colour of an indicator changes over a certain pH range, sometimes called 'colour change interval' or 'sensitivity range' and within this pH range there is a definite pH for each indicator at which it shows its *neutral colour*. It will be clear from the following table.

Indicator	pH *range of colour change*	*Acid colour*	*Alakaline colour*	*Neutral colour* pH
Methyl orange	3·1–4·4	Red	Orange yellow	4·0
Methyl red	4·2–6·3	Red	Yellow	5·0
Litmus	5·0–8·0	Red	Blue	7·0
Phenolphthalein	8·2–10·0	Colourless	Pink	8·5
Thymolphthalein	9·3–10·5	Colourless	Blue	10·0

During titration of acids with alkalies having different strengths, the choice of indicators may be made from the table given below.

Titrations	*Suitable indicators*	pH *of colour change*
Strong acid and strong base	Any	pH 5·0 to 9·0
Strong acid and weak base	Methyl orange	pH 4·0
Weak acid and strong base	Phenolphthalein	pH 8·5
Weak acid and weak base	None satisfactory	Slow change of pH

Standard solutions and their units

A standard solution of a substance contains a known weight of it in a definite volume of the solution. Hence a standard solution is a solution of known strength or concentration. There are various types of standard solutions which are generally used in laboratories. The units of these standard solutions are mentioned below.

Percentage concentration

(*a*) Weight percentage or percent by weight or % W/W: This is defined as the number of parts of solutes by weight per 100 parts of solution by weight. For example, a 2% solution of sodium chloride contains 2 gm of sodium chloride in 98 gm of distilled water.

(*b*) Volume percentage or percent by volume or % V/V: In case of a solution of liquid in liquid it is better to express percentage by volume which means the number of parts of a liquid substance per 100 parts of the substance by volume. Thus 5% solution of alcohol contains 5 ml absolute alcohol in 95 ml of distilled water.

For ordinary purpose we generally prepare percentage solution in terms of W/V (weight by volume) which ignores the volume changes of the solvent with temperature (1 ml of water weighs 1 gm at 4°C.).

Molar and Molal solutions (M)

(*a*) Molarity or volume molar solution: A molar solution contains 1 gm molecule of a substance per litre of the solution at 20°C. Thus a molar solution of H_2SO_4 contains 98 gm of it per litre and a molar solution of NaOH contains 40 gm of it per litre solution.

(*b*) MOLALITY OR WEIGHT MOLAR SOLUTION : A molal solution contains 1 gm molecule of the substance per 1000 gm of the solvent. Thus a 5 molal solution of NaOH contains (40×5) or 200 gm of it in 1000 gm of water. This is independent of changes in temperature.

NORMAL SOLUTION (N)

A normal solution of a substance contains 1 gm equivalent (equivalent weight expressed in grams) of it per litre of the solution. Equivalent weight of a substance may be determined as follows.

$$\text{Equivalent weight} = \frac{\text{Molecular weight}}{\text{H or OH per mol}}.$$

The following are the equivalent weights of chemicals which are commonly used :

HCl—36·46 ;	H_2SO_4—49·04 ;	HNO_3—63 ;
CH_3COOH—60·05 ;	$COOH.COOH,2H_2O$—63 ;	H_3PO_4—32·67 ;
$NaOH$—40 ;	KOH—56 ;	$Ca(OH)_2$—37 ;
NH_3—17 ;	NH_4OH—35 ;	Na_2CO_3—53 ;
$CaCO_3$—50 ;	$FeCl_3, 6H_2O$—90·17 ;	$AgNO_3$—170 ;
$AgCl$—143·34 ;	$CuSO_4,5H_2O$—124·75 ;	$Ba(OH)_2$—85·67 ;
$KMnO_4$—31·60 ;	$K_2Cr_2O_7$—49 ;	$Na_2S_2O_3,5H_2O$—248.

GRAM PER LITRE : This is expressed as the weight of a solute in gram per litre of the solution.

PARTS PER MILLION (ppm) : The term ppm means parts (in grams) present per million millilitre of the solution. Thus to prepare a ppm solution of substance weigh the substance in milligram (if the desired concentration is *x*; then *x* mg) and make the volume up to one litre with the solvent. This is also expressed as mg/litre or μg/ml.

To prepare ppm solution of IAA, GA_3 and kinetin, the required amount of the hormone in mg is first dissolved in 1-2 ml of 0·05 M (0·2%) NaOH or absolute alcohol and then make the volume up to a litre. Abscisic acid is highly soluble in NaOH, other inhibitors are mostly soluble in water, e.g., CCC, AMO-1618, B-995, etc.

CONVERSION OF MOLAR STRENGTH INTO NORMAL STRENGTH

$$\text{Gram equivalent weight of a salt} = \frac{\text{Gram molecular weight}}{\text{No. of metal atoms} \times \text{its valency}}.$$

$$\text{Molarity} = \frac{\text{Mol. wt.} \times \text{Molar strength}}{\text{Equivalent wt.}} \text{Normality.}$$

Thus gram molecular weight of 1 M NaOH=40.

$$\text{Therefore its gram equivalent} = \frac{40 \times 1}{1 \times 1} = 40.$$

Hence 1 M NaOH=1 N NaOH.

$$\text{Similarly 1 M } Na_2CO_3 = \frac{106 \times 1}{2 \times 1} \text{ or } \frac{106}{53} \text{ or 2 N } Na_2CO_3.$$

Conversion of a percent solution into normal solution

$$\text{Normality} = \frac{10 \times \text{percent strength}}{\text{Gram equivalent}}.$$

$$\text{Normality of a 8\% NaOH} = \frac{10 \times 8}{40} \text{ or } 2 \text{ N}.$$

Conversion of gram per litre solution into normality

$$\text{Normality} = \frac{\text{Weight of solute in gram per litre}}{\text{Gram equivalent}}.$$

Thus a solution of NaOH of strength 60 gm/litre has a normality of 60/40 or 1·5 N.

Concentrations of some commonly used acids and NH_4OH (liquid)

	Acetic acid (*Glacial* A.R.)	HCl (A.R.)	H_2SO_4 (A.R.)	HNO_3 (A.R.)	NH_4OH (A.R.)
Molecular wt.	60·03	36·46	98·08	63·02	35·05
Specific gravity	1·05	1·18	1·84	1·42	0·90
Normality of conc. reagent	17·40	11·60	36·00	16·00	15·00
Approx. percent present in conc. reagent.	99·5	37·00	96·00	70·00	58·00 (29·00 NH_3)
No. of ml. required to dilute to one litre to make 6N reagent.	350	516	168	390	400

If the purity and density of acids are known then the following formula may be applied to prepare a desired strength in normality.

$$\text{Volume of conc. acid} = \frac{100 \times \text{equivalent wt.} \times \text{desired strength}}{\text{purity} \times \text{density}} \text{ ml.}$$

This volume should be made up to a litre with distilled water to have the desired strength. For example, to prepare a 0·5 N HCl, the purity of which is 35·39 and density 1·18, the volume of the conc. acid is

$$V = \frac{100 \times 36{\cdot}5 \times 0{\cdot}5}{35{\cdot}39 \times 1{\cdot}18} \text{ ml.}$$

This volume should be diluted with water to make one litre to get 0·5 N HCl.

Approximate figures for concentrated acids, bases and ethanol

	Mol. wt.	*%by weight*	gm/litre	*Molarity*	*Millilitre required for 1000* ml N *soln.*
HCl	36·47	37·5(1·19)	445	12·2	82·0
HNO_3	63·02	70·0(1·41)	989	15·7	63·8
H_2SO_4	98·08	95·0(1·84)	1742	17·8	28·2
H_3PO_4	98·04	86·0(1·70)	1462	14·9	22·4
CH_3COOH	60·03	99·5(1·05)	1406	17·4	57·4
NH_4OH (as NH_3)	17·03	28·0(0·90)	252	14·8	67·6
NaOH soln.	40·01	50·0(1·53)	763	19·1	52·5
KOH soln.	56·11	52·0(1·54)	800	14·3	70·2
C_2H_5OH (95%)	46·05	92·3(0·816)	753	16·4	61·2
C_2H_5OH (Abs.)	46·05	99·1(0·796)	789	17·1	58·4

Approximate density (grams/ml) is given within parenthesis.

To prepare x**% solution from** y**% stock solution:** x ml of the stock solution is to be diluted with the solvent up to y ml, i.e., $(y-x)$ ml solvent is to be added to x ml of the stock solution. Thus to prepare 5% solution of NaOH from a 40% stock solution, 5 ml of the stock solution is to be diluted with (40−5) ml or 35 ml water. This is applicable to all units of standard solution.

Variation of density of water with temperature

Temp. in °C.	0	2	4	6	8
0	0·99987	0·99997	1·0000	0·99997	0·99980
10	0·99973	0·99953	0·99927	0·99897	0·99862
20	0·99823	0·99780	0·99732	0·99681	0·99626
30	0·99567	0·99505	0·99440	0·99371	0·99300
40	0·99220	0·99150	0·99070	0·98980	0·98900
50	0·98810	0·98720	0·98620	0·98530	0·98430

Warburg's manometer fluid

Anhydrous NaBr ... 44 gm
Triton -*x* or Teepol ... 300 ml
Evan's blue ... 300 mg
Distilled water ... 1000 ml

Lanolin paste of auxin : The required amount of auxin is dissolved in 1-2 ml of ethanol and then thoroughly mixed with pure lanolin assuming that 1 gm of lanolin equals to 1 ml of water. For example, for a 1000 ppm or 0·1% paste 100 mg of auxin is dissolved in 1–2 ml ethanol and the solution is mixed thoroughly with 100 gm of lanolin.

Some useful formulae

Area of a circle $= \pi r^2$
Area of a cylinder $= \pi r^2 h$
Volume of a sphere $= \frac{4}{3}\ \pi r^2$
Area of an elipse $= (\pi/4)\ (a \times b)$
Volume of a cone $= \frac{1}{3}$ area of base $\times$ height.

Some common constants

$\pi = 3{\cdot}1416$
$g = 32$ ft. per sec. per sec. $= 981$ cm per sec. per sec.
1 ml of water at 4°C. = 1 gm
Density of air at N.T.P. $= 0{\cdot}00129$/ml
1 litre of hydrogen at N.T.P. = 22·4 litres
$e = 2{\cdot}7183$ $\log_e = 0{\cdot}4343$.
Gas constant R = 0·082 litre atmosphere per degree per mole
Planck's constant $= 6{\cdot}63 \times 10^{-27}$ erg sec.

Some important isotopes and their physical characters

Isotopes	*Mass number*	*Half-life*	*Particles emitted*
Arsenic (As)	76 & 77	26·7 *h* & 39 *h*	β & γ
Barium (Ba)	131 & 133	12 *d* & 7·5 *y*	K & γ
Carbon (C)	14	5600 *y*	β
Calcium (Ca)	41, 45 & 47	10^5 *y*, 160 *d* & 4·7 *d*	K; β & β, γ
Chlorine (Cl)	36	3×10^5 *y*	β
Cadmium (Cd)	115	43 *d*	β & γ
Cobalt (Co)	58 & 60	71 *d* & 5·27 *y*	β^+, γ & β, γ
Iron (Fe)	55 & 59	2·9 *y* & 45 *d*	K & β, γ
Hydrogen (H)	3	12.3 *y*	β
Iodine (I)	129 & 131	10^7 *y* & 8·05 *d*	β^+, γ & β, γ
Potassium (K)	40 & 42	10^9 *y* & 12.4 *h*	β^+, K, γ & β, γ
Sodium (Na)	12 & 24	2·6 *y* & 15 *h*	β^+, K, γ & β, γ
Phosphorus (P)	32	14·2 *d*	β
Lead (Pb)	210 & 202	19·4 *y* & 10^5 *y*	β, γ, e & L
Rubidium (Rb)	86	18.6 *h*	β, γ
Sulphur (S)	35	87 *d*	β
Tritium (Ti)	204	3·56 *y*	β, K
Uranium (U)	238, 234, 235 & 233	$4{\cdot}5 \times 10^9$ *y*, $2{\cdot}5 \times 10^5$ *y* $7{\cdot}1 \times 10^8$ *y* & $1{\cdot}6 \times 10^5$ *y*	α, γ
Zinc (Zn)	95 & 93	65 *d* & 9×10^5 *y*	β, γ, e & β, γ

Table of Atomic weights

Name	*Symbol*	*Atomic number*	*Atomic weight*
Aluminium	Al	13	26·98
Antimony	Sb	51	121·75
Arsenic	As	33	74·92
Barium	Ba	56	137·34
Boron	B	5	10·81
Bromine	Br	35	79·90
Cadmium	Cd	48	112·40
Calcium	Ca	20	40·08
Carbon	C	6	12·01
Chlorine	Cl	17	35·45
Chromium	Cr	24	51·99
Cobalt	Co	27	58·93
Copper	Cu	29	63·54
Gold	Au	79	196·96
Hydrogen	H	1	1·008
Iodine	I	53	126·90
Lead	Pb	82	207·19
Magnesium	Mg	12	24·312
Manganese	Mn	25	54·93
Mercury	Hg	80	200·59
Molybdenum	Mo	42	95·94
Nickel	Ni	28	58·71
Nitrogen	N	7	14·00
Oxygen	O	8	15·999
Phosphorus	P	15	30·97
Potassium	K	19	39·10
Rubidium	Rb	37	85·47
Silicon	Si	14	28·08
Silver	Ag	47	107·87
Sodium	Na	11	22·989
Strontium	Sr	38	87·62
Sulphur	S	16	32·06
Zinc	Zn	30	65·37

Control of relative humidity

(a) BY SULPHURIC ACID MIXTURE METHOD

Percent H_2SO_4	*Percentage Relative humidity*	*Vapour pressure*
None	100·0	17·39
1·57	99·5	17·30
3·03	99·1	17·20
4·49	98·7	17·20
5·96	98·2	17·10
7·37	97·5	17·00

Percent H_2SO_4	Percentage Relative humidity	Vapour pressure
8·77	96·90	16·90
10·19	96·20	16·70
11·60	95·60	16·60
12·99	94·80	16·50
14·35	93·90	16·30
15·71	93·20	16·20
17·01	92·30	16·10
18·31	91·20	15·90
19·61	89·90	15·60
20·91	88·80	15·40
22·19	87·40	15·20
23·47	85·70	14·90
24·76	84·00	14·60
26·04	82·30	14·30
27·32	80·50	14.00
28·58	78·70	13·70
29·84	76·70	13·30
31·11	74·60	13·00
32·28	72·50	12·60
33·43	70·40	12·20
34·57	68·00	11·80
35·71	65·50	11·40
36·87	63·10	11·00
38·03	60·70	10·60
39·19	58·30	10·10
44·00	49·00	8·50
46·00	45·00	7·70
48·00	42·00	7·10
50·00	38·00	6·50
52·00	33·00	5·80
54·00	29·50	5·00
56·00	25·00	4·30
58·00	21·50	3·50
60·00	18·50	3·00
62·00	15·50	2·60
64·00	12·70	2·20
66·00	10·50	1·80
68·00	9·00	1·50
70·00	7·50	1·30
72·00	6·00	1·00
74·00	4·50	0·60
76·00	3·50	0·50
78·00	3·00	0·40
80·00	2·50	0·30
82·00	1·50	0·20

(b) BY SATURATED SALT SOLUTIONS

Salt	*Temp.* °F.	*Percentage Relative humidity*
$BaCl_2, 2H_2O$	85	88
$CaCl_2$	20	44
	32	41
	50	40
	70	35
$CaCl_2, 6H_2O$	41	39·8
	68	32·3
	76	31
$CaSO_4, 5H_2O$	68	98
$Ca(NO_3)_2$	20	64
	32	64
	50	59
	70	55
$Ca(NO_3)_2, 4H_2O$	68	53·6
	77	50·4
	86	46·6
	95	42
	100	38·9
KBr	68	84
$K_2CO_3, 2H_2O$	66	44
	76	43
$KC_2H_3O_2$ (K-acetate)	68	23·2
	77	22·7
	86	22
	100	20·4
KNO_2	68	49
	77	48·2
	86	47·2
	100	45·9
KNO_3	32	97·6
	50	95·5
	68	93·2
	86	90·7
	104	87·9
K_2CrO_4	68	86·6
	77	86·5
	100	85·6
K_2SO_4	32	99·1
	50	97·9
	68	97·2
	104	96·2
	122	95·8
$LiCl, H_2O$	32	14·7
	68	12·4
	86	11·8

Salt	*Temp. °F.*	*Percentage Relative humidity*
LiCl, H_2O	104	11·6
	122	11·4
LiCl	68	12·2
	77	11·2
	86	11·2
	100	11·2
$MgCl_2$	73	32·9
	86	32·4
	100	31·4
$MgCl_2$, $6H_2O$	32	35
	68	33·6
	86	32·8
	104	32·1
	122	31·4
NaCl	32	74·9
	68	75·5
	86	75·6
	104	75·4
	122	74·7
NH_4Cl	22	82
	32	83
	50	81
	70	75

In the range from 32°F. to 122°F., the deviation from average relati humidity is 0 to ±0·6% varying from one salt to another.

Conversion of some commonly used units and dimensions

1 Acre = 43,560 sq.ft. = 4,047 sq. metres
1 Angstrom = 3,937 × 10^{-9} inches = 1 × 10^{-10} metres = 1 × 10^{-4} micron(
1 Atmosphere = 76·0 cm of Hg
1 Bar = 0·9869 atmosphere
1° Centigrade (Celsius) = (1 × $\frac{9}{5}$) + 32°F. = $\frac{5}{4}$ Remer
1 Centimetre = 0·3937 inches = 0·01 metre
1 Cubic centimetre = 10^{-6} cubic metres = 0·001 litre
1 Day = 86,400 seconds = 1440 minutes
1 Degree (angular) = 0·01745 radian
1 Dyne/sq. cm. = 9·869 × 10^{-7} atmospheres
1° Farenheit = (1 − 32) × $\frac{5}{9}$°C.
1 Foot = 30·48 centimetres = 0·3048 metre

1 Foot-candle =10·764 lumen/sq. metre

1 Gallor =3·785 litres

1 Gram =980·7 dynes =0·03527 ounces (Avdp.) =0·03215 (Troy) $=2 \cdot 205 \times 10^{-3}$ pounds

1 Gram/cubic centimetre =62·43 pounds/cu. ft.

1 Gram/litre =1000 parts (in mg) per million or 1000 ppm

1 Hour $=4 \cdot 167 \times 10^{-2}$ days

1 Hectare =2·47 acres =10,000 sq. metres

1 Inch =2·540 centimetres $=2 \cdot 540 \times 10^{-2}$ metres

1 Lumen/sq. ft. =1 foot-candle =10·764 lumen/sq. metre =0·07958 spherical candle power

1 Lux =0·0929 foot-candles

1 Metre =1·09 yards

1 Microgram $=10^{-6}$ grams

1 Microlitre $=10^{-6}$ litres

1 Micron $=10^{-3}$ mm $=10^{-4}$ cm $=10^{-6}$ m

1 Mile =0·6214 kilometres

1 Millimicron $=10^{-6}$ mm $=10^{-7}$ cm $=10^{-9}$ m

1 Milligram/litre =1 ppm

1 ml = 1·000027 cu. cm. (c.c.)

1 Ounce =28·349527 grams

1 Ounce (fluid) =0·02957 litre

1 Pound =453·5924 grams

1 Quintal =100 kilograms =220·46 pounds =0·1 tonne

1 Radian =57·30 degrees

1 Square centimetre $=1 \cdot 076 \times 10^{-3}$ sq. ft. =0·0001 sq. metres

1 Yard =91·44 centimetres

N.B. The basic units (SI) of length, volume and mass are metre, litre, and gram respectively. The following prefixes may be used to denote decimal fractions or multiples of the basic SI units:

Prefix	*Symbol*	*Fraction*	*Prefix*	*Symbol*	*Multiple*
deci	d	10^{-1}	deca	da	10
centi	c	10^{-2}	hecto	h	10^{2}
milli	m	10^{-3}	kilo	k	10^{3}
micro	μ	10^{-6}	mega	M	10^{6}
nano	n	10^{-9}	giga	G	10^{9}
pico	p	10^{-12}	tera	T	10^{12}
femto	f	10^{-15}			
atto	a	10^{-18}			

The following are some of the latest terminologies which are quickly finding usage in International System of Units (SI).

Commonly used	*Modern usage (SI)*
ångström (Å)	nanometre (1 Å $= 10^{-1}$ nm)
atmosphere (atm)	kilopascal (1 atm $\approx$ 101 kPa)
bar	kPa (1·0 bar = 100 kPa)
billiön	10^9 or 10^{12} *
calorie (cal)	joule (1 cal $\approx$ 4·2 J)
centigrade	Celsius (°C)
counts per minute (c.p.m.)	ct/min
cubic centimetre	cm^3 or ml for general purpose
dextrose	glucose (G *or* Glc)
dialysate	diffusate
disintegrations per minute (d.p.m.)	d/min
dry ice	solid CO_2
dyne (dyn)	micro newton (1 dyn=10 μN)
equivalent (unit of substance)	mole
erg (erg)	micro joule (1 erg=0·1 μJ)
ethyl alcohol	ethanol
foot	metre (1 ft $\approx$ 0·3 m)
gallon (gal)	litre (1 gal $\approx$ 4·5 litres)
gamma (γ)	microgram (μg)
gram-molecule	mole
half-life	$T_{\frac{1}{2}}$
isozyme	isoenzyme
Krebs cycle	tricarboxylic acid cycle
laevulose	fructose
lambda (λ)	microlitre (μl) or mm^3
lux	lx
methyl alcohol	methanol
micromicro-($\mu\mu$)	pico-
micron (μ)	micrometre (μm)
millimicro-(10^{-9})	nano-
millimicron (mμ)	nanometre (nm)
molar (conc.)	mol/l
N. T. P. (normal temperature and pressure)	s. t. p. (standard temperature and pressure)
parts per million (p. p. m.)	parts/10^6
parts per thousand	parts/10^3
petroleum ether	light petroleum (with b. p. range)
pound	kilogram (1lb=0·45 kg)
revolutions per minute (r.p.m.)	rev/min
specific gravity (sp. gr.)	relative density
weight by volume (w/v)	mass/vol. (m/v)

*** 10^9 or a thousand millions used in France and U.S.A., also called milliard; 10^{12} or a million millions used in Britain.**

Conversion of centrifugal force (g) into r.p.m.

The value of G (gravitational force) for each speed and rotor is required to know frequently in physiological procedures, such as separation of different intra-cellular components at a given gravitational force. The speed of the centrifuge can be determined in r.p.m. with the help of an instrument called *tachometer*. G can be calculated from the equation given below. G is the gravitational force exerted on the homogenate being centrifuged ; defined as the gravitational force exerted on one gram mass at a distance r (cm) from the axis of rotation.

$G = \frac{S^2 \times r}{89{,}500}$, where G = relative gravitational force in multiple of g (gravitational constant 980 cm/sec.^{2}).

r = radial distance in cm from the centre or axis of rotation.

S = revolution per minute or r.p.m.

or $G = (1{\cdot}118 \times 10^{-5}) \times (\text{r.p.m.})^2 \times r$

If $G = 7{,}000 \times g$

$r = 10$ cm

$$S = \sqrt{\frac{89{,}500 \times 7{,}000}{10}}$$

= 7900 r.p.m.

Use of Logarithms for calculations (use of five-figure tables)

Logarithms: The logarithm of a number consists of an integral part called *characteristic*, or *index*, and a decimal part, the *mantissa*. Referring to the logarithmic table it will be seen at rows of five-figures are placed against each of the numbers from 10 to 99. These five figures form in each case the mantissa of logarithm ; the characteristic has to be supplied in each case.

The characteristic of any number greater than unity is positive, and is less by one than the number of figures to the left of the decimal point. The characteristic of a number less than unity is negative and is greater by one than the number of zeros which follow the decimal point.

Characteristic of	73727 is 4	characteristic	of	737·27 is 2
,,	,, 73·727 is 1	,,	,,	7·3727 is 0
,,	,, 0·73727 is $\bar{1}$	,,	,,	0·073727 is $\bar{2}$
,,	,, 0·0073727 is $\bar{3}$	and so on.		

Negative characteristics are usually designated as *bar* 1, *bar* 2, etc.

To find the logarithm of a given number :

Example 1 : To find log 7·3.

Here the characteristic is 0 and in the column opposite the number 73 in the log-table is found the mantissa 86332.

Hence, log 7·3 =0·86332.

Similarly, log 73 =1·86332 ; log 7300 =3·86332.

To obtain the logarithm of three figures, first find the first two numbers at the extreme left ; then passing along the horizontal line to the number in the vertical column headed by the third figure, the value of logarithm of three figures is obtained.

EXAMPLE 2 : To find log 737.

log 737 =2·86747.

To obtain the logarithm of a number consisting of four or five figures, it is necessary to use the *mean difference* columns at the extreme right of the page.

EXAMPLE 3 : To find log 737·2

Mantissa of log 737	=	0·86747
Mean diff. for 2	=	12
∴ log 737·2	=	2·86759

EXAMPLE 4 : To find log 73·727

Mantissa of log 737	=	0·86747	
Mean diff. for 2	=	12	
Mean diff. for 7	=	4	1
log 73·727	=	1·86763	

Similarly, log 0·0073727 $=\bar{3}\cdot86763$.

Antilogarithms : The number corresponding to a given logarithm is found by using the logarithmic table for antilogarithms.

EXAMPLE 5 : To find the number whose log is 1·59584

Antilog of ·595	=	0·39355	
Mean diff. for 8	=	72	
Mean diff. for 4	=	3	6
		0·39431	

Hence the number whose log is 1·59584 is 39·431. Similarly the number whose log is $\bar{3}\cdot59584$ is 0·0039431.

POSITIVE AND NEGATIVE CHARACTERISTICS : It is observed that the characteristic of a logarithm may be negative. In log $\bar{3}\cdot59584$ only the $\bar{3}$ (bar 3) is negative, the remaining five figures are positive. The positive and negative characteristics used to put the point of decimal as will be clear from the examples.

Addition :

EXAMPLE 6 : To add 3·30535 and $\bar{4}$·54654.

Sum $=\bar{1}$·85189 ; Sum characteristics 3 plus $\bar{4}=\bar{1}$.

Subtraction :

EXAMPLE 7 : To subtract $\bar{2}$·87506 from $\bar{3}$·74036.

To subtract, change the signs of the characteristic of the former and add. Thus, 1 carried from the mantissa gives $\bar{4}$, and $\bar{4}$ plus 2 $=\bar{2}$.

$$\begin{array}{ll} \text{Therefore,} & \bar{3}\cdot74036 \\ & 2\cdot87506 \\ \hline & \bar{2}\cdot86530 \end{array}$$

Multiplication : Add the logarithms of the numbers and the sum is the logarithm of the product.

EXAMPLE 8 : Multiply 37·358 by 0·0058343

$$\begin{array}{ll} \log\ 37.358 & =1\cdot57238 \\ \log 0\cdot0058343 & =\bar{3}\cdot76599 \\ \hline \text{Sum} & =\bar{1}\cdot33837 \quad \therefore \text{Product} = 0\cdot21796. \end{array}$$

Division : Subtract the logarithm of division from the logarithm of the dividend ; the difference is the logarithm of the quotient.

EXAMPLE 9 : Divide 0·37358 by 25·687.

$$\begin{array}{ll} \log\ 0\cdot37358 & =\bar{1}\cdot57238 \\ \log 25\cdot687 & =1\cdot40972 \\ \hline \text{Difference} & =\bar{2}\cdot16266 \quad \therefore \text{Quotient} = 0\cdot014543. \end{array}$$

Involution : Multiply the logarithm of the given number by the index of the power ; the product is the logarithm of the required number.

EXAMPLE 10 : To find the cube of 0·36715.

$$\log (0\cdot36715)^3 = 3\times\log 0\cdot36715 = 3\times\bar{1}\cdot56485 = \bar{2}\cdot69455.$$
$$\text{Antilog} = 0\cdot049494.$$

Here $3\times\bar{1}=\bar{3}$, but 1 carried from the preceding figure gives $\bar{2}$.

LOGARITHMS

	0	1	2	3	4	5	6	7	8	9	Mean Differences.								
											1	2	3	4	5	6	7	8	9
10	00000	00432	00860	01284	01703						42	85	127	170	212	254	297	339	381
						02119	02531	02938	03342	03743	40	81	121	162	202	242	283	323	364
11	04139	04532	04922	05308	05690						37	77	116	154	193	232	270	309	348
						06070	06446	06819	07188	07555	37	74	111	148	185	222	259	296	333
12	07918	08279	08636	08991	09342						36	71	106	142	177	213	248	284	319
						09691	10037	10380	10721	11059	34	68	102	136	170	204	238	272	307
13	11394	11727	12057	12385	12710						33	66	98	131	164	197	229	262	295
						13033	13354	13672	13988	14301	32	63	95	126	158	190	221	253	284
14	14613	14922	15229	15534	15836						30	61	91	122	152	183	213	244	274
						16137	16435	16732	17026	17319	29	59	88	118	147	177	206	236	265
15	17609	17898	18184	18469	18752						28	57	85	114	142	171	199	228	256
						19033	19312	19590	19866	20140	28	55	83	110	138	165	193	221	248
16	20412	20683	20951	21219	21484						27	53	80	107	134	160	187	214	240
						21748	22011	22272	22531	22789	26	52	78	104	130	156	182	208	233
17	23045	23300	23553	23805	24055						26	50	76	101	126	151	176	201	227
						24304	24551	24797	25042	25285	25	49	73	98	122	147	171	196	220
18	25527	25768	26007	26245	26482						24	48	71	95	119	143	167	190	214
						26717	26951	27184	27416	27646	23	46	69	93	116	139	162	185	208
19	27875	28103	28330	28556	28780						23	45	68	90	113	135	158	180	203
						29003	29226	29447	29667	29885	22	44	66	88	110	132	154	176	198
20	30103	30320	30535	30750	30963	31175	31387	31597	31806	32015	21	43	64	85	106	127	148	170	190
21	32222	32428	32634	32838	33041	33244	33445	33646	33846	34044	20	41	61	81	101	121	141	162	182
22	34242	34439	34635	34830	35025	35218	35411	35603	35793	35984	20	39	58	77	97	116	135	154	174
23	36173	36361	36549	36736	36922	37107	37291	37475	37658	37840	19	37	56	74	93	111	130	148	167
24	38021	38202	38382	38561	38739	38917	39094	39270	39445	39620	18	35	53	71	89	106	124	142	159

LOGARITHMS

	0	1	2	3	4	5	6	7	8	9	Mean Differences. 1	2	3	4	5	6	7	8	9
25	39794	39967	40140	40312	40483	40654	40824	40993	41162	41330	17	34	51	68	85	102	119	136	153
26	41497	41664	41830	41996	42160	42325	42488	42651	42813	42975	16	33	49	65	82	98	115	131	148
27	43136	43297	43457	43616	43775	43933	44091	44248	44404	44560	16	32	47	63	79	95	111	126	142
28	44716	44871	45025	45179	45332	45484	45637	45788	45939	46090	15	30	46	61	76	91	107	122	137
29	46240	46389	46538	46687	46835	46982	47129	47276	47422	47567	15	29	44	59	74	88	103	118	132
30	47712	47857	48001	48144	48287	48430	48572	48714	48855	48996	14	29	43	57	72	86	100	114	129
31	49136	49276	49415	49554	49693	49831	49969	50106	50243	50379	14	28	41	55	69	83	97	110	124
32	50515	50650	50786	50920	51054	51188	51322	51455	51587	51720	13	27	40	54	67	80	94	107	121
33	51851	51983	52114	52244	52375	52504	52634	52763	52892	53020	13	26	39	52	65	78	91	104	117
34	53148	53275	53403	53529	53656	53782	53908	54033	54158	54283	13	25	38	50	63	76	88	101	113
35	54407	54531	54654	54777	54900	55023	55145	55267	55388	55509	12	24	37	49	61	73	85	98	110
36	55630	55751	55871	55991	56110	56229	56348	56467	56585	56703	12	24	36	48	60	71	83	95	107
37	56820	56937	57054	57171	57287	57403	57519	57634	57749	57864	12	23	35	46	58	70	81	93	104
38	57978	58092	58206	58320	58433	58546	58659	58771	58883	58995	11	23	34	45	57	68	79	90	102
39	59106	59218	59329	59439	59550	59660	59770	59879	59988	60097	11	22	33	44	55	66	77	88	99
40	60206	60314	60423	60531	60638	60746	60853	60959	61066	61172	11	21	32	43	54	64	75	86	97
41	61278	61384	61490	61595	61700	61805	61909	62014	62118	62221	10	21	31	42	53	63	74	84	95
42	62325	62428	62531	62634	62737	62839	62941	63043	63144	63246	10	20	31	41	51	61	71	82	92
43	63347	63448	63548	63649	63749	63849	63949	64048	64147	64246	10	20	30	40	50	60	70	80	90
44	64345	64444	64542	64640	64738	64836	64933	65031	65128	65225	10	20	29	39	49	59	68	78	88
45	65321	65418	65514	65610	65706	65801	65896	65992	66087	66181	10	19	29	38	48	57	67	76	86
46	66276	66370	66464	66558	66652	66745	66839	66932	67025	67117	9	19	28	37	47	56	65	74	84
47	67210	67302	67394	67486	67578	67669	67761	67852	67943	68034	9	18	27	36	46	55	64	73	82
48	68124	68215	68305	68395	68485	68574	68664	68753	68842	68931	9	18	27	36	45	53	63	72	81
49	69020	69108	69197	69285	69373	69461	69548	69636	69723	69810	9	18	26	35	44	53	62	70	79

LOGARITHMS

	0	1	2	3	4	5	6	7	8	9	Mean Differences. 1	2	3	4	5	6	7	8	9
50	69897	69984	70070	70157	70243	70329	70415	70501	70586	70672	9	17	26	34	43	52	60	69	77
51	70757	70842	70927	71012	71096	71181	71265	71349	71433	71517	8	17	25	34	42	50	59	67	76
52	71600	71684	71767	71850	71933	72016	72099	72181	72263	72346	8	17	25	33	42	50	58	66	75
53	72428	72509	72591	72673	72754	72835	72916	72997	73078	73159	8	16	24	32	41	49	57	65	73
54	73239	73320	73400	73480	73560	73640	73719	73799	73878	73957	8	16	24	32	40	48	56	64	72
55	74036	74115	74194	74273	74351	74429	74507	74586	74663	74741	8	16	23	31	39	47	55	63	70
56	74819	74896	74974	75051	75128	75205	75282	75358	75435	75511	8	15	23	31	39	46	54	62	69
57	75587	75664	75740	75815	75891	75967	76042	76118	76193	76268	8	15	23	30	38	45	53	60	68
58	76343	76418	76492	76567	76641	76716	76790	76864	76938	77012	7	15	22	30	37	44	52	59	67
59	77085	77159	77232	77305	77379	77452	77525	77597	77670	77743	7	15	22	29	37	44	51	58	66
60	77815	77887	77960	78032	78104	78176	78247	78319	78390	78462	7	14	22	29	36	43	50	58	65
61	78533	78604	78675	78746	78817	78888	78958	79029	79099	79169	7	14	21	28	36	43	50	57	64
62	79239	79309	79379	79449	79518	79588	79657	79727	79796	79865	7	14	21	28	35	41	48	55	62
63	79934	80003	80072	80140	80209	80277	80346	80414	80482	80550	7	14	20	27	34	41	48	54	61
64	80618	80686	80754	80821	80889	80956	81023	81090	81158	81224	7	13	20	27	34	40	47	54	60
65	81291	81358	81425	81491	81558	81624	81690	81757	81823	81889	7	13	20	26	33	40	46	53	59
66	81954	82020	82086	82151	82217	82282	82347	82413	82478	82543	7	13	20	26	33	39	46	52	59
67	82607	82672	82737	82802	82866	82930	82995	83059	83123	83187	6	13	19	26	32	38	45	51	58
68	83251	83315	83378	83442	83506	83569	83632	83696	83759	83822	6	13	19	25	32	38	44	50	57
69	83885	83948	84011	84073	84136	84198	84261	84323	84386	84448	6	12	19	25	31	37	43	50	56
70	84510	84572	84634	84696	84757	84819	84880	84942	85003	85065	6	12	19	25	31	37	43	50	56
71	85126	85187	85248	85309	85370	85431	85491	85552	85612	85673	6	12	18	24	31	37	43	49	55
72	85733	85794	85854	85914	85974	86034	86094	86153	86213	86273	6	12	18	24	30	36	42	48	54
73	86332	86392	86451	86510	86570	86629	86688	86747	86806	86864	6	12	18	24	30	35	41	47	53
74	86923	86982	87040	87099	87157	87216	87274	87332	87390	87448	6	12	17	23	29	35	41	46	52

LOGARITHMS

	0	1	2	3	4	5	6	7	8	9	Mean Differences 1	2	3	4	5	6	7	8	9
75	87506	87564	87622	87679	87737	87795	87852	87910	87967	88024	6	12	17	23	29	35	41	46	52
76	88081	88138	88195	88252	88309	88366	88423	88480	88536	88593	6	11	17	23	29	34	40	46	51
77	88649	88705	88762	88818	88874	88930	88986	89042	89098	89154	6	11	17	22	28	34	39	45	50
78	89209	89265	89321	89376	89432	89487	89542	89597	89653	89708	6	11	17	22	28	33	39	44	50
79	89763	89818	89873	89927	89982	90037	90091	90146	90200	90255	6	11	17	22	28	33	39	44	50
80	90309	90363	90417	90472	90526	90580	90634	90687	90741	90795	5	11	16	22	27	32	38	43	49
81	90848	90902	90956	91009	91062	91116	91169	91222	91275	91328	5	11	16	21	27	32	37	42	48
82	91381	91434	91487	91540	91593	91645	91698	91751	91803	91855	5	11	16	21	27	32	37	42	48
83	91908	91960	92012	92064	92117	92169	92221	92273	92324	92376	5	10	16	21	26	31	36	42	47
84	92428	92480	92531	92583	92634	92686	92737	92788	92840	92891	5	10	15	20	26	31	36	41	46
85	92942	92993	93044	93095	93146	93197	93247	93298	93349	93399	5	10	15	20	26	31	36	41	46
86	93450	93500	93551	93601	93651	93702	93752	93802	93852	93902	5	10	15	20	25	30	35	40	45
87	93952	94002	94052	94101	94151	94201	94250	94300	94349	94399	5	10	15	20	25	30	35	40	45
88	94448	94498	94547	94596	94645	94694	94743	94792	94841	94890	5	10	15	20	25	29	34	39	44
89	94939	94988	95036	95085	95134	95182	95231	95279	95328	95376	5	10	15	19	24	29	34	39	44
90	95424	95472	95521	95569	95617	95665	95713	95761	95809	95856	5	10	14	19	24	29	34	38	43
91	95904	95952	95999	96047	96095	96142	96190	96237	96284	96332	5	9	14	19	24	28	33	38	42
92	96379	96426	96473	96520	96567	96614	96661	96708	96755	96802	5	9	14	19	24	28	33	38	42
93	96848	96895	96942	96988	97035	97081	97128	97174	97220	97267	5	9	14	18	23	28	32	38	42
94	97313	97359	97405	97451	97497	97543	97589	97635	97681	97727	5	9	14	18	23	28	32	37	42
95	97772	97818	97864	97909	97955	98000	98046	98091	98137	98182	5	9	14	18	23	27	32	36	41
96	98227	98272	98318	98363	98408	98453	98498	98543	98588	98632	5	9	14	18	23	27	32	36	41
97	98677	98722	98767	98811	98856	98900	98945	98989	99034	99078	4	9	13	18	22	27	31	36	40
98	99123	99167	99211	99255	99300	99344	99388	99432	99476	99520	4	9	13	18	22	26	31	35	40
99	99564	99607	99651	99695	99739	99782	99826	99870	99913	99957	4	9	13	17	22	26	31	35	39

ANTILOGARITHMS

	0	1	2	3	4	5	6	7	8	9	Mean Differences. 1	2	3	4	5	6	7	8	9
·00	10000	10023	10046	10069	10093	10116	10139	10162	10186	10209	2	5	7	9	12	14	16	19	21
·01	10233	10257	10280	10304	10328	10351	10375	10399	10423	10447	2	5	7	10	12	14	17	19	21
·02	10471	10495	10520	10544	10568	10593	10617	10641	10666	10691	2	5	7	10	12	15	17	20	22
·03	10715	10740	10765	10789	10814	10839	10864	10889	10914	10940	3	5	8	10	13	15	18	20	23
·04	10965	10990	11015	11041	11066	11092	11117	11143	11169	11194	3	5	8	10	13	15	18	20	23
·05	11220	11246	11272	11298	11324	11350	11376	11402	11429	11455	3	5	8	11	13	16	18	21	24
·06	11482	11508	11535	11561	11588	11614	11641	11668	11695	11722	3	5	8	11	13	16	19	21	24
·07	11749	11776	11803	11830	11858	11885	11912	11940	11967	11995	3	5	8	11	14	16	19	22	25
·08	12023	12050	12078	12106	12134	12162	12190	12218	12246	12274	3	6	8	11	14	17	20	22	25
·09	12303	12331	12359	12388	12417	12445	12474	12503	12531	12560	3	6	9	11	14	17	20	23	26
·10	12589	12618	12647	12677	12706	12735	12764	12794	12823	12853	3	6	9	12	15	18	21	24	26
·11	12882	12912	12942	12972	13002	13032	13062	13092	13122	13152	3	6	9	12	15	18	21	24	27
·12	13183	13213	13243	13274	13305	13335	13366	13397	13428	13459	3	6	9	12	15	18	21	25	28
·13	13490	13521	13552	13583	13614	13646	13677	13709	13740	13772	3	6	9	13	16	19	22	25	28
·14	13804	13836	13868	13900	13932	13964	13996	14028	14060	14093	3	6	10	13	16	19	22	25	29
·15	14125	14158	14191	14223	14256	14289	14322	14355	14388	14421	3	7	10	13	16	20	23	26	30
·16	14454	14488	14521	14555	14588	14622	14655	14689	14723	14757	3	7	10	13	17	20	24	27	30
·17	14791	14825	14859	14894	14928	14962	14997	15031	15066	15101	3	7	10	14	17	21	24	28	31
·18	15136	15171	15205	15241	15276	15311	15346	15382	15417	15453	4	7	11	14	18	21	25	28	32
·19	15488	15524	15560	15596	15631	15668	15704	15740	15776	15812	4	7	11	14	18	22	25	29	32
·20	15849	15885	15922	15959	15996	16032	16069	16106	16144	16181	4	7	11	15	18	22	26	30	33
·21	16218	16255	16293	16331	16368	16406	16444	16482	16520	16558	4	8	11	15	19	23	26	30	34
·22	16596	16634	16672	16711	16749	16788	16827	16866	16904	16943	4	8	12	15	19	23	27	31	35
·23	16982	17022	17061	17100	17140	17179	17219	17258	17298	17338	4	8	12	16	20	24	28	32	36
·24	17378	17418	17458	17498	17539	17579	17620	17660	17701	17742	4	8	12	16	20	24	28	32	36

ANTILOGARITHMS.

	0	1	2	3	4	5	6	7	8	9	Mean Differences.								
											1	2	3	4	5	6	7	8	9
·25	17783	17824	17865	17906	17947	17989	18030	18072	18113	18155	4	8	12	17	21	25	29	33	37
·26	18197	18239	18281	18323	18365	18408	18450	18493	18535	18578	4	8	13	17	21	25	30	34	38
·27	18621	18664	18707	18750	18793	18836	18880	18923	18967	19011	4	9	13	17	22	26	30	35	39
·28	19055	19099	19143	19187	19231	19275	19320	19364	19409	19454	4	9	13	18	22	26	31	35	40
·29	19498	19543	19588	19634	19679	19724	19770	19815	19861	19907	5	9	14	18	23	27	32	36	41
·30	19953	19999	20045	20091	20137	20184	20230	20277	20324	20370	5	9	14	19	23	28	32	37	42
·31	20417	20464	20512	20559	20606	20654	20701	20749	20797	20845	5	10	14	19	24	29	33	38	43
·32	20893	20941	20989	21038	21086	21135	21184	21232	21281	21330	5	10	15	19	24	29	34	39	44
·33	21380	21429	21478	21528	21577	21627	21677	21727	21777	21827	5	10	15	20	25	30	35	40	45
·34	21878	21928	21979	22029	22080	22131	22182	22233	22284	22336	5	10	15	20	25	31	36	41	46
·35	22387	22439	22491	22542	22594	22646	22699	22751	22803	22856	5	10	16	21	26	31	37	42	47
·36	22909	22961	23014	23067	23121	23174	23227	23281	23336	23388	5	11	16	21	27	32	37	43	48
·37	23442	23496	23550	23605	23659	23714	23768	23823	23878	23933	5	11	16	22	27	33	38	44	49
·38	23988	24044	24099	24155	24210	24266	24322	24378	24434	24491	6	11	17	22	28	34	39	45	50
·39	24547	24604	24660	24717	24774	24831	24889	24946	25003	25061	6	11	17	23	29	34	40	46	51
·40	25119	25177	25236	25293	25351	25410	25468	25527	25586	25645	6	12	18	23	29	35	41	47	53
·41	25704	25763	25823	25882	25942	26002	26062	26122	26182	26242	6	12	18	24	30	36	42	48	54
·42	26303	26363	26424	26485	26546	26607	26669	26730	26792	26853	6	12	18	24	31	37	43	49	55
·43	26915	26977	27040	27102	27164	27227	27290	27353	27416	27479	6	13	19	25	31	38	44	50	56
·44	27542	27606	27669	27733	27797	27861	27925	27990	28054	28119	6	13	19	26	32	39	45	51	58
·45	28184	28249	28314	28379	28445	28510	28576	28642	28708	28774	7	13	20	26	33	39	46	52	59
·46	28840	28907	28973	29040	29107	29174	29242	29309	29376	29444	7	13	20	27	34	40	47	54	60
·47	29512	29580	29648	29717	29785	29854	29923	29992	30061	30130	7	14	21	28	34	41	48	55	62
·48	30200	30269	30339	30409	30479	30549	30620	30690	30761	30832	7	14	21	28	35	42	49	56	63
·49	30903	30974	31046	31117	31189	31261	31333	31405	31477	31550	7	14	22	29	36	43	50	58	65

ANTILOGARITHMS.

	0	1	2	3	4	5	6	7	8	9	Mean Differences. 1	2	3	4	5	6	7	8	9
·50	31623	31696	31769	31842	31916	31989	32063	32137	32211	32285	7	15	22	29	37	44	52	59	66
·51	32359	32434	32509	32584	32659	32735	32809	32885	32961	33037	8	15	23	30	38	45	53	60	68
·52	33113	33189	33266	33343	33420	33497	33574	33651	33729	33806	8	15	23	31	39	46	54	62	69
·53	33884	33963	34041	34119	34198	34277	34356	34435	34514	34594	8	16	24	32	40	47	55	63	71
·54	34674	34754	34834	34914	34995	35075	35156	35237	35318	35400	8	16	24	32	40	48	56	65	73
·55	35481	35563	35645	35727	35810	35892	35975	36058	36141	36224	8	16	25	33	41	50	58	66	74
·56	36308	36392	36475	36559	36644	36728	36813	36898	36983	37068	8	17	25	34	42	51	59	68	76
·57	37154	37239	37325	37411	37497	37584	37670	37757	37844	37931	9	17	26	35	43	52	61	69	78
·58	38019	38107	38194	38282	38371	38459	38548	38637	38726	38815	9	18	27	35	44	53	62	71	80
·59	38905	38994	39084	39174	39264	39355	39446	39537	39628	39719	9	18	27	36	45	54	63	72	82
·60	39811	39902	39994	40087	40179	40272	40365	40458	40551	40644	9	19	28	37	46	56	65	74	83
·61	40738	40832	40926	41020	41115	41210	41305	41400	41495	41591	9	19	28	38	47	57	66	76	85
·62	41687	41783	41879	41976	42073	42170	42267	42364	42462	42560	10	19	29	39	49	58	68	78	87
·63	42658	42756	42855	42954	43053	43152	43251	43351	43451	43551	10	20	30	40	50	60	70	80	89
·64	43652	43752	43853	43954	44055	44157	44259	44361	44463	44566	10	20	30	41	51	61	71	81	91
·65	44668	44771	44875	44978	45082	45186	45290	45394	45499	45604	10	21	31	42	52	62	73	83	94
·66	45709	45814	45920	46026	46132	46238	46345	46452	46559	46666	11	21	32	43	53	64	75	85	96
·67	46774	46881	46989	47098	47206	47315	47424	47534	47643	47753	11	22	33	44	54	65	76	87	98
·68	47863	47973	48084	48195	48306	48417	48529	48641	48753	48865	11	22	33	45	56	67	78	89	100
·69	48978	49091	49204	49317	49431	49545	49659	49774	49888	50003	11	23	34	46	57	68	80	91	103
·70	50119	50234	50350	50466	50582	50699	50816	50933	51050	51168	12	23	35	47	58	70	82	93	105
·71	51286	51404	51523	51642	51761	51880	52000	52119	52240	52360	12	24	36	48	60	72	84	96	108
·72	52481	52602	52723	52845	52966	53088	53211	53333	53456	53580	12	24	37	49	61	73	85	98	110
·73	53703	53827	53951	54075	54200	54325	54450	54576	54702	54828	13	25	38	50	63	75	88	100	113
·74	54954	55081	55208	55336	55463	55590	55719	55847	55976	56105	13	26	38	51	64	77	90	102	115

ANTILOGARITHMS

	0	1	2	3	4	5	6	7	8	9	Mean Differences. 1	2	3	4	5	6	7	8	9
·75	56234	56364	56494	56624	56754	56885	57016	57148	57280	57412	13	26	39	52	66	79	92	105	118
·76	57544	57677	57810	57943	58076	58210	58345	58479	58614	58749	13	27	40	54	67	80	94	107	121
·77	58884	59020	59156	59293	59429	59566	59704	59841	59979	60117	14	27	41	55	69	82	96	110	123
·78	60256	60395	60534	60674	60814	60954	61094	61235	61376	61518	14	28	42	56	70	84	98	112	126
·79	61659	61802	61944	62087	62230	62373	62517	62661	62806	62951	14	29	43	58	72	86	101	115	130
·80	63096	63241	63387	63533	63680	63826	63973	64121	64269	64417	15	29	44	59	74	88	103	118	132
·81	64565	64714	64863	65013	65163	65313	65464	65615	65766	65917	15	30	45	60	75	90	105	120	135
·82	66069	66222	66374	66527	66681	66834	66988	67143	67298	67453	15	31	46	62	77	92	108	123	139
·83	67608	67764	67920	68077	68234	68391	68549	68707	68865	69024	16	32	47	63	79	95	110	126	142
·84	69183	69343	69503	69663	69823	69984	70146	70307	70469	70632	16	32	48	64	81	97	113	129	145
·85	70795	70958	71121	71285	71450	71614	71779	71945	72111	72277	17	33	50	66	83	99	116	132	149
·86	72444	72611	72778	72946	73114	73282	73451	73621	73790	73961	17	34	51	68	85	101	118	135	152
·87	74131	74302	74473	74645	74817	74989	75162	75336	75509	75683	17	35	52	69	87	104	121	138	156
·88	75858	76033	76208	76384	76560	76736	76913	77090	77268	77446	18	35	53	71	89	107	125	142	159
·89	77625	77804	77983	78163	78343	78524	78705	78886	79068	79250	18	36	54	72	91	109	127	145	163
·90	79433	79616	79799	79983	80168	80353	80538	80724	80910	81096	19	37	56	74	93	111	130	148	167
·91	81283	81470	81658	81846	82035	82224	82414	82604	82794	82985	19	38	57	76	95	113	132	151	170
·92	83176	83368	83560	83753	83946	84140	84333	84528	84723	84918	19	39	58	78	97	116	136	155	175
·93	85114	85310	85507	85704	85901	86099	86298	86497	86696	86896	20	40	60	79	99	119	139	158	178
·94	87096	87297	87498	87700	87902	88105	88308	88512	88716	88920	20	41	61	81	102	122	142	162	183
·95	89125	89331	89536	89743	89950	90157	90365	90573	90782	90991	21	42	62	83	104	125	146	166	187
·96	91201	91411	91622	91833	92045	92257	92470	92683	92897	93111	21	42	64	85	106	127	149	170	191
·97	93325	93541	93756	93972	94189	94406	94624	94842	95060	95280	22	43	65	87	109	130	152	174	195
·98	95499	95719	95940	96161	96383	96605	96828	97051	97275	97499	22	44	67	89	111	133	155	178	200
·99	97724	97949	98175	98401	98628	98855	99083	99312	99541	99770	23	46	68	91	114	137	160	182	205

APPENDIX III

SOME INSTRUCTIONS TO STUDENTS

Students of Plant Physiology should be more alert and cautious while attending practical classes. They are also expected to have some aptitude for handling precision apparatus and equipments to perform any experiment with them. It should be borne in mind that unless the students observe strict discipline in the class it is not possible for the teacher or the instructor to help the individual student. It should also be remembered that in the practical class the students are required to learn how to perform a particular experiment on some physiological process and its modification by some treatments. One might not think that by performing the experiment once in the class within the limited period, one should get the correct result. Deviations from the correct or the expected results may always take place with living organs or organisms. It is all important to find out some reasonable explanations to these variations. This of course needs thinking power and a good knowledge about the subject.

Some general instructions are given below and if these are always kept in mind and followed rigorously while doing practical classes, the students will be immensely benefited in the long run.

1. Always go to your laboratory with inquisitiveness. First read your instructions carefully. In case you find any difficulty don't hesitate to consult your teacher for help. Careful planning of your work will not only enable you to perform the experiment more elegantly but also minimise your time and effort.

2. Report all your experiments honestly even though the results may appear to be wrong. In that case repeat the experiment if time permits. Think yourself and try to find out the possible sources of error in your experiments and how these can be eliminated. Remember that good results can only be achieved if one is critical enough in each and every step of the experiment. For statistical calculations several replications are necessary, even though the results may appear to be correct.

3. Maintain cleanliness ; keep your glass wares clean. Do not contaminate the reagents by indiscriminate use of pipettes, spatula, etc., or by interchanging stoppers of reagent bottles. All reagent bottles should be kept in their respective positions being stoppered properly after use. Use just as much chemicals or reagents as are necessary for the experiment.

4. Keep all inflammable solvents and reagents away from flame. Do not heat any inflammable solvent directly over a flame, it will at once catch fire and may bring about some unfortunate incident. Always use water bath for such purposes. Do not point test tube mouths at another person or yourself while heating over a flame. While diluting acids, do not pour water into acids but slowly add conc. acids into water. Do not heat measuring cylinders, volumetric flasks or reagent bottles over a flame.

5. Report all breakages of glass wares, glass equipments and any damage of apparatus to either your teacher or the laboratory attendant.

6. While using analytical balance, take utmost care. Do not put the materials to be weighed directly on the pan, use oiled paper, or polythene sheet or watch glass or weighing bottle for the purpose. Place the sample

on the *left* pan and weights on the *right*. Always use forceps for transfer of weights taking special care in case of rider and small weights. Analytical balance is a very sensitive apparatus and handle it carefully.

7. While using any precision apparatus or sophisticated instrument first learn the correct procedure for use from your teacher or instructor and then operate it taking sufficient precaution and care as per directions. Rough handling of such apparatus must be avoided.

8. Discuss your experiment and compare your results with your fellow students. You will learn a lot and become aware of the range of variations in the biological materials.

9. Last and not the least, students should develop a feeling of love and affection for all the belongings of their laboratory, not to speak of apparatus and equipments only.

Report of experimental data

After the completion of each experiment, a written report is to be furnished to the teacher or the instructor concerned. This report should be prepared on a ruled laboratory note book, the left-hand unruled pages are to be used for drawing figures or attaching graph sheets. The laboratory note book should bear the student's name and roll number and sessional year. The report should be submitted regularly and got checked and signed by the teacher or the instructor.

The report should be represented as follows :

1. *Title of the experiment.*
2. *Date of experiment.*
3. *Object of the experiment :* State very briefly the main purpose of the experiment.
4. *Requirements for the experiment :* This may be categorised as follows :
   (a) Plant material.
   (b) Chemicals or reagents required.
   (c) Glass wares and equipments.
   (d) Miscellaneous.

5. EXPERIMENTAL METHODS OR PROCEDURE : State in short the general procedure adopted for the experiment and do not write anything which is not pertinent to the experiment. Suitable figures of equipments or apparatus used for the purpose may be given wherever necessary.

6. OBSERVATIONS AND RESULTS : Represent your data in tables and graphs. Properly label the axes of graphs and give a suitable caption for each table and graph. If possible record the room temperature, relative humidity, light intensity, etc. For rapid calculation of results always use logarithmic table.

7. COMMENT AND DISCUSSION : Write in a few sentences about the conclusion of your experiment. If there is any variation from the expected results, try to find out the possible sources of error and write them categorically with reasoning. Always remember that biological processes are influenced by external and internal environmental factors. Interpret the data with the background of your knowledge in the subject and finally write the concluding remarks.

